HANDBOOK OF QUALITY ASSURANCE FOR THE ANALYTICAL CHEMISTRY LABORATORY

HANDBOOK OF QUALITY ASSURANCE FOR THE ANALYTICAL CHEMISTRY LABORATORY

James P. Dux, Ph.D.

VAN NOSTRAND REINHO

Library of Congress Catalog Card Number: 85-6145
ISBN: 0-442-21972-5

Manufactured in the United States of America

Published by Van Nostrand Reinhold Company Inc.
115 Fifth Avenue
New York, New York 10003

Van Nostrand Reinhold Company Limited
Molly Millars Lane
Wokingham, Berkshire RG11 2PY, England

Van Nostrand Reinhold
480 La Trobe Street
Melbourne, Victoria 3000, Australia

Macmillan of Canada
Division of Canada Publishing Corporation
Boulevard
Canada

7 6 5 4 3

in Publication Data

ce for the analytical

ality control. I. Title.
28 85-6145

To my wife, Catherine,
and our children Thomas and Ann

PREFACE

Over forty years ago, when the author was first employed as an analytical chemist, the term "quality assurance" or "quality control" as applied to chemical analysis would have been met with blank stares. The overwhelming majority of analyses in those days were run by the "wet chemical" quantitative methods of gravimetry and titrimetry. The only analytical instruments widely available were filter colorimeters, and colorimetric methods were looked upon with suspicion and used only when classical methods could not be conveniently applied. After all, it was widely known that colorimetric methods had an accuracy of only $\pm 5\%$, about ten times less accurate than the classical methods.

Because of the emphasis on classical wet chemical techniques, all laboratory workers were degreed chemists, since it was felt that only through the four-year chemistry curriculum would the required knowledge and skill be acquired. Technicians or non-degreed laboratory workers were simply not hired, except for menial tasks, e.g. washing glassware, running errands, or making coffee.

As a result of the use of highly precise methods, applied by professional chemists, laboratory data was seldom questioned. If it was, a simple rerun of the sample sufficed to confirm the original result, or to demonstrate that something had gone wrong. Statistical analysis of data was primitive, as illustrated by the fact that the "average deviation" of replicate analyses (usually triplicate) was the customary method for establishing precision.

The contrast with the modern analytical laboratory is remarkable. In today's lab the majority of analyses are run by instrumental methods, often by non-degreed "technicians." The analytical chemist today is a supervisor, or a specialist in one or more of the more esoteric instrumental techniques which have not yet been reduced to the routine level at which the technician can operate. In addition, of course, the analytical chemist must be available for interpretation of the data generated in the lab.

Two factors have caused this transformation. Most obvious is the development over the years of more sophisticated instrumentation, generated in large part by the revolution in electronic technology and by the availability of superior materials of construction. Less obvious, but possibly more important, has been the need to increase productivity in the lab, i.e., the number of analyses run per dollar of labor spent. The old classical techniques were highly labor intensive, while the new instrumental techniques are capital intensive. Over the years the cost of labor has risen much more rapidly than the cost of capital, making the investment in productivity improvement through instrumentation very attractive.

An auxiliary benefit of the improved instrumentation was a steady drive toward lower detection limits. The old classical methods had limits of detectability generally of the order of 0.001% or 10 parts per million. Modern instrumentation can detect certain contaminants at parts per billion levels, and, using special techniques in certain cases, parts per trillion. The result has been a great increase of interest in the extent of environmental pollution, contamination of food supplies, workplace industrial hygiene, and similar concerns to the point where "chemophobia" or fear or chemicals has become widespread in the population. It is not generally recognized that the basis for these concerns is the improvement in analytical techniques which has come about in the last two or three decades.

The result, of course, has been increased pressure on industry from government regulatory agencies, consumer groups, and organized labor to reduce as far as possible the pollution of the external environment, the workplace environment, and the food supply. Billions of dollars have been and are being spent to accomplish these objectives.

These factors, namely the increased use of non-professional personnel, the increasing complexity of instrumentation, the extremely low detection limits (with concomitant possibility of contamination), and most importantly, the fact that large amounts of money are being spent based on analytical data, have combined to focus an increasing concern with the "quality" of laboratory generated data. A widely publicized scandal involving allegedly fraudulent laboratory data led to the Food and Drug Administration's publication of "Good Laboratory Practices" or GLP's, which were also adopted by the Environmental Protection Agency. Professional analytical chemists themselves, especially those who have been involved with interlaboratory methods validation studies, have become aware of the alarming variability of results from different laboratories analyzing the same samples by (supposedly) the same methods.

These developments have led to great pressure on laboratory supervisors and managers to establish quality assurance programs to maintain control of quality and to document quality assurance procedures. Unfortunately, quality assurance, as it applies to analytical chemistry laboratories, is not well defined in the technical literature, nor is it commonly taught as part of an analytical chemist's education. Most of the literature is fragmented, often contradictory, and tends toward the theoretical. This book is intended to serve as a handbook for laboratory supervisors so that they may design a quality assurance program which will serve the needs of their managers and clients, and be acceptable to their technical personnel.

JAMES P. DUX, PH.D.

ACKNOWLEDGMENTS

I should like to acknowledge my friends and colleagues at Lancaster Laboratories, Inc. for their many helpful discussions and suggestions which have gone into this book, and express my thanks to Senior Management at the Laboratories for permission to reprint many of the figures used.

CONTENTS

HANDBOOK OF QUALITY ASSURANCE FOR THE ANALYTICAL CHEMISTRY LABORATORY

1
QUALITY CONTROL AND QUALITY ASSURANCE

In the past few years the words "quality," "quality control," and "quality assurance" have almost achieved the status of bywords in the American vocabulary. This has been due mainly to the success of our Japanese competitors in today's worldwide marketplace, especially in the fields of cameras, electronic consumer goods, and automobiles. The Japanese have taken the twin concepts and quality and reasonable prices as marketing basics and succeeded in dominating these fields of manufacture.

What is perhaps not so widely known is that the concepts and techniques of quality control exploited by the Japanese had their origin in the first three decades of this century in the United States. Industrial engineers in those years established the statistical principles necessary to control and assure quality in assembly line manufacturing operations. These techniques found widespread application during World War II when quality production was mandatory due to a workforce which was hastily trained and often using new, innovative processes.

Superficially, it is tempting to compare the analytical laboratory to a manufacturing process and assume that the same principles of quality control and assurance apply. The samples received by the lab may be thought of as raw material, upon which a series of operations are performed, resulting in a "product" which is an analytical report. Many of the techniques used for quality control and assurance in the laboratory, e.g., control charts and equipment calibration, are similar in form to those used in manufacturing. However, the analogy cannot be stretched too far. For example, raw material samples received by the laboratory are not uniform, and the "product" is not a tangible object, but rather *information* about the raw material which can be used in many different ways. The techniques used to control and assure quality in laboratory data are similar to those used in manufacturing, but are subtly different in application and interpretation. In addition most laboratory operations resemble those of the job shop more than the assembly line.

The terms "quality," "quality control," and "quality assurance" are frequently used very loosely in the analytical chemistry literature. It is assumed that "quality" of analytical data is a concept which is intuitively understood and need not be defined, when, in actuality, different meanings may be intended by the speaker or writer than the listener or reader has in mind. It is instructive to give some thought to the meaning of the term "quality" as applied to the analytical laboratory. In general, there are three levels of interpretation, which should be kept in mind when discussing quality control or quality assurance.

At the most elementary level we are talking about the quality of the data itself. The quality of a single datum i.e., an analytical result, may be simply defined as its accuracy, i.e., the degree to which it approaches the "true" value of the concentration of the analyte being determined. In fact, the quality of a single analytical result could be defined mathematically as:

$$Q = 1 - | X - T | / T \tag{1-1}$$

where X is the result and T is the true value. If $X = T$, $Q = 1$ signifying perfect accuracy or highest quality. For $X = 0$, or $X = 2T$, $Q = 0$, signifying complete lack of quality. This would also be true for negative Q which would result if $X > 2T$. The problem, of course, is that in most analyses T is not known and so the quality of an individual analytical result is not known.

At the second level we encounter the term quality in connection with analytical methods. The quality of a method is assumed to be a function of the degree to which it is free of systematic error, or bias, and the degree to which it is free of random errors. The first of these parameters, i.e., freedom from bias, is often called the "accuracy" of the method, and the second is called the precision. However, note that the term "accuracy" used in this sense is somewhat different than the term accuracy as defined above, where it is taken to mean the closeness to the truth, regardless of whether the difference between X and

T is due to systematic or random error. In other words, a method with no bias, but a large random error may give a result with far less accuracy than a method with a small bias and small random variation. The logical absurdity of calling the first method which may yield less accurate results more accurate than the second which yields more accurate results is apparent.[1] In this book we will confine the term accuracy to the first definition, i.e., the degree to which *X* approaches *T*.

Another semantic difficulty arises with the term precision. Usually the precision is defined in terms of the standard deviation or the 95% or 99% confidence limits derived from the standard deviation. However, the more precise a method is, the smaller is the standard deviation. In other words the precision is an inverse function of the standard deviation, which is in actuality, a measure of the *imprecision* of the method.

Much of the technical analytical literature is devoted to this aspect of evaluating the quality of analytical methodology. When a new method is developed it is tested by an experienced analyst, using the best instrumentation, under perfect conditions. Results are then evaluated to determine the degree of freedom from bias and imprecision. While this type of evaluation is useful in comparing one method to another, it has little bearing on the quality (accuracy) of results obtained when the method is used on a daily basis, with less skilled analysts, and all of the possibilities for human error that exist in the real world.

This leads us to the third level of meaning for the word quality, i.e., the quality of the analytical system, where the latter encompasses all of the components which can affect the outcome of the analysis, both human and inanimate. Two laboratories with the same equipment, and with personnel with equal training and education, may differ widely in the quality (accuracy) of the data they turn out. The difference is due to a difference in supervision or management, in other words, in the organization for quality control and the amount of manpower expended in quality control. This level of meaning of the word quality is the subject of this book.

QUALITY CONTROL

Quality control, in the sense in which it is used in this book, may be defined as those operations undertaken in the laboratory to ensure that the data produced are generated within known probability limits of accuracy and precision. Note that nothing is said about the actual level of accuracy and precision, only that they are known to a certain level of confidence. The level of quality produced in a laboratory in a given analysis may be a function of many considerations, e.g., convenience, availability of equipment, safety, speed, and above all, cost. It is generally of no value to generate data with a degree of accuracy far greater than the need in the use to which the data is going to be put. If a machine is producing several thousand pounds of product per hour and a control analysis takes several hours to generate a result, several thousand pounds of useless product may have been made while waiting for the analysis. It would be far better to run a quicker test with a lower quality method, provided the quality were good enough for the purpose of control. The important point is that the laboratory supervisor should know the limits of accuracy of his data, and this is the objective of quality control.

QUALITY ASSURANCE

Although quality control represents the core activity in a quality assurance program, the latter encompasses much more than the technical operations of controlling quality. Quality assurance consists of the system whereby the laboratory can assure clients and other outside investigators, e.g., government agencies, accrediting bodies, etc., that the laboratory is generating data of proven and known quality. Quality assurance depends primarily on documentation, which is designed with the following objectives:

(a) To demonstrate that the quality control operations are, in fact, being carried out. For example, quality control may dictate that a pH meter be calibrated against standard buffer solutions each day it is used. Quality assurance dictates that the analyst write down in his notebook that the pH meter was calibrated that day, the results obtained, and any indication that the meter was not functioning correctly.

(b) To assure accountability of the data, i.e., that the data reported do in fact represent analyses on the sample as submitted or collected. In other words, that there are safeguards against sample mix-up.

(c) To assure traceability of reported data. Each result reported should be easily traceable to the analyst who ran the test, the method used, the raw data collected, the instrument used and its condition, and the status of the quality control system at the time the sample was run.

(d) To demonstrate that reasonable precautions are being taken against the possibility of falsification of data, i.e., that the data cannot easily be tampered with at some future date. An example would be the use of bound notebooks rather than looseleaf or spiral bound notebooks.

QUALITY ASSURANCE AND LABORATORY MANAGEMENT

It should be evident that establishing and maintaining a good quality assurance program is a management task and must represent a commitment on the part of management which is communicated from the top down. Analytical laboratories today serve many functions in our society. Some are part of a manufacturing organization, some are independent testing laboratories, others are part of R & D organizations or attached to government agencies. Quality assurance is important to all of them, but sometimes it is hard to convince senior management, especially management with budgetary control, of the necessity for a good program. The reason for this is that quality assurance represents a pure cost and is difficult to justify in terms of the proverbial "bottom line." In this sense it is analogous to a good safety program, in that it can only be justified in terms of disasters averted. The consequences of poor quality data can be worse than no data at all, since it may lead to a false confidence in analytical results, or lead to action which can in turn result in a hazardous or costly blunder. "Assurance" is equivalent to "insurance" and as such represents a relatively inexpensive safeguard against such mistakes. Senior management must make it plain to all subordinates that it supports a good quality assurance program and will underwrite the cost.

Laboratory workers may often balk at the installation of a QA program such as described in this book. One reason for this is that it represents a new way of doing things and it is human nature to resist change. Another is that it represents more work for the same amount of output, and much of the work is simply designed to assure others that the lab is doing a good job, which the workers already know they are doing. Finally, it generates a considerable amount of paperwork which must be read, evaluated, filed, and acted upon. Laboratory workers are oriented by education and disposition toward working with "things," i.e., instruments and other apparatus, and resent having to spend time on paperwork.

Nevertheless, it has been the author's experience that a good QA program, once it is in place and working, invariably results in an increase in morale and pride in the work. One reason for this is that the program provides a constant reinforcement of the feeling that the work is being done properly. With a good QA program the analyst not only knows the quality of his results, but also has the data to prove it. The ultimate vindication of the QA program comes when the analyst must defend his results to a client or some other outside agency, or even in a court of law.

REFERENCES

1. Kirchmer, C. J., Quality Control in Water Analysis, *Environmental Science and Technology,* 17: 174A–184A (1983).

2
ELEMENTARY STATISTICS

It is generally recognized that any type of measurement is subject to the possibility of error. Indeed if a measurement is made which yields a result coinciding exactly with the "true" or expected value of the parameter being measured, the knowledgeable scientist will regard it as fortuitous, or due to ignoring the capability of the measuring device, i.e., not measuring to the customary number of "significant figures" or digits. By training, the scientist or engineer is taught to push the capability of his instrument to the limit by reading gauges, calibration marks, and charts to the smallest "uncertain" digit. Thus, even if there were no inherent errors in the instrument or apparatus, or in the chemical manipulations preceding the final measurement, there will always be the possibility of error in the final estimate. In today's world of digital read-out of instruments, the task of determining the least significant digit is left to the instrument, but it is no less uncertain.

Errors in measurement are usually classified into two broad categories. The first of these is "assignable" errors, often called "bias," i.e., errors which are always in the same direction, either positive or negative, usually of the same magnitude, and which can be traced to an assignable cause, e.g., a miscalibration of an instrument or failure to correct for a reagent blank. In general, bias in an analytical method can be detected by running samples of known concentration, for example, and corrected by changing the operational procedures in the method or in the calculation of the final result.

The other broad category of errors is "random error." These are the errors which cannot be assigned to any known cause and are due to random variations in the way indicating devices are read. It is in the measurement and control of these random errors that the techniques of statistical evaluation of data are most applicable.

It is essential that the analyst concerned with quality control be familiar with, and comfortable with, the application of statistics to the generation and evaluation of data. It is not essential that the analyst understand the underlying derivation of the mathematical formulae involved, but it is important to understand the implications of the numbers generated. In this chapter we will ignore the derivation of equations and confine ourselves to their application. Fortunately, the mathematical manipulations seldom involve more than elementary algebra.

If a large, homogeneous sample of material is subjected to repeated application of an analytical method, say *n* times, *n* results will be obtained. The *n* results will occupy a range defined by the smallest and largest values. If *n* is sufficiently large, it will generally cluster around some value approximately midway in the range. In other words, there will be fewer values the further away we get from the midpoint of the range.

This tendency to cluster and the degree of spread of the results are two important parameters of the data. The clustering or *central tendency* is most often measured by the arithmetic mean, or average, of the data. This is simply the sum of all of the data, divided by the number *n:*

$$\bar{X} = (X_1 + X_2 + \cdot \cdot \cdot + X_n)/n = (\Sigma X_i)/n \qquad (2\text{–}1)$$

Most people intuitively recognize the average value as being somehow a "better" estimate of the true value than any of the individual numbers which make it up. There are other measures of the central tendency, such as the *median,* or mid-value, which is the number which divides the group in half (half being larger and half smaller), or the *mode* which is the number which occurs most often. However, for analytical chemistry quality control, the average is the most useful measure; and if the variation in results is truly random, the mean, median, and mode will coincide.

The dispersion, or spread of the data, can also be measured in many ways. One way is the range, as defined above. The range can be very useful, but obviously is very sensitive to both unusually high or unusually low results.

The most useful measure of dispersion for our purposes, as for most statistical operations, is the *standard deviation*. The standard deviation, or root mean square deviation, is defined mathematically as:

$$s = [\Sigma (X_i - \overline{X})^2/(n - 1)]^{1/2} \qquad (2\text{–}2)$$

Another useful measure of dispersion often used in applied statistics is the variance, which is simply the square of the standard deviation:

$$v = s^2 = [\Sigma (X_i - \overline{X})^2/(n - 1)] \qquad (2\text{–}3)$$

To return to our group of n determinations, let us assume that n is very large, say greater than 100. If these n data are plotted in the form of a frequency distribution, i.e., a plot of the number of times a certain value is obtained against the value itself, the resulting curve will approximate Figure 2–1, which is known by a variety of names, e.g., the *normal* curve, *normal distribution, Gaussian* curve or distribution, *bell-shaped* curve, etc.

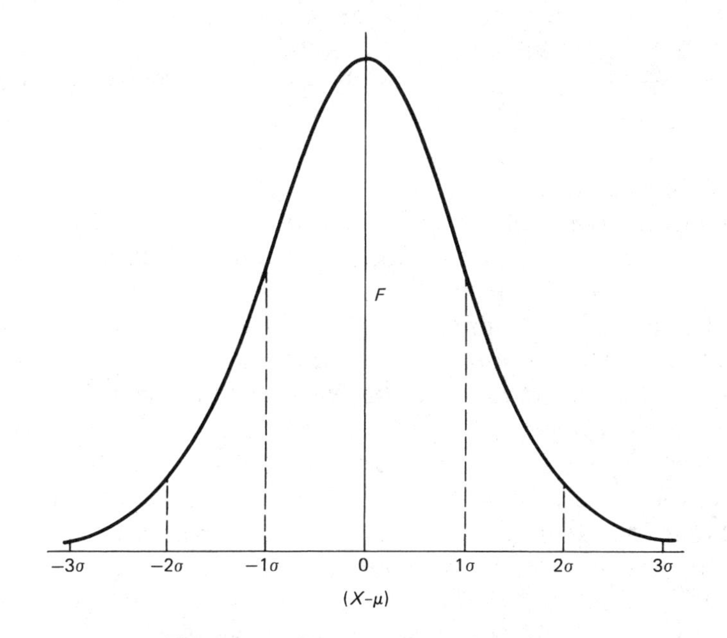

Figure 2–1 The Normal Distribution

The importance of this distribution is that many natural phenomena, e.g., the results of chemical analyses, will give rise to this type of curve, if the underlying causes of variation are truly random, and if the probability of a variation occurring is inversely proportional to its magnitude. If the variation is truly random, positive variation will occur as often as negative. If the variation is inversely proportional to its magnitude, large variations will occur less frequently than small.

The normal distribution is a curve, as shown in Figure 2-1, which can be described mathematically in terms of the two parameters: μ and σ. The symbol μ refers to the mean of the universe, or population, of all possible results, or in other words, the average of an infinite number of analytical determinations; in our case, if $n = \infty$. The symbol σ refers to the correspondingly calculated standard deviation. The equation of the curve is given by:

$$F(X) = (1/(2\pi)^{1/2}\sigma)\exp[-(X-\mu)^2/2\sigma^2)] \qquad (2\text{–}4)$$

where $F(X)$ represents the frequency with which the value X is obtained.

It can be shown by straightforward integration techniques that 68.26% of the area under the curve falls within one standard deviation of the mean, i.e., within $\pm\sigma$. Similarly, 95.46% of the area falls within $\pm 2\sigma$, and 99.73% falls within $\pm 3\sigma$. The significance of this fact is that any *single* determination has a 68.26% probability of being within one standard deviation of the mean, a 95.46% probability of being within two standard deviations of the mean, etc.

What is more important is that, if we know σ, but not μ and run a single analysis, we can state that the mean, or true value, has a 95% probability of being, for example, within $\pm$ 2 σ of the result we have obtained. Hence, is born the concept of the *confidence limits,* i.e., a range within which we can state, with some numerical probability, that the *true* value of an analysis lies. For example, if σ is known to be $\pm$0.04 ppm, and a single analysis yields a result of 0.80 ppm, we can state, with 95% confidence that we are correct, that the true concentration of the analyte is 0.8 $\pm$ .08 ppm ($\pm$ 2 σ) or between 0.72 and 0.88 ppm; and with 99.7% confidence that it is 0.8 $\pm$ 0.12 ($\pm$ 3 σ), or between 0.68 and 0.92. The meaning of the 95% confidence limit is that if we make such statements 100 times, we would only be wrong 5 times, and at the 99% confidence level we would be wrong only once in 100 times.

PRACTICAL APPLICATIONS OF THE STANDARD DEVIATION

Unfortunately, we seldom have the time or inclination to run the very large number of analyses needed to calculate σ, the standard deviation of the universe or population of all possible outcomes of the analysis. This dilemma was very neatly addressed by the chemist W. S. Gossett, who published a means for calculating confidence limits, or probabilities, for finite groups of analytical determinations. Gossett, writing under the name "Student" (said to be because he was fearful of his employer learning that he was wasting his time working on statistical theory) introduced the concept of the *t*-factor, which is a function of the desired probability level and the number of analyses run, and by which the standard deviation s (as calculated in equation 2–2 above) must be multiplied to yield the desired confidence limits. Table 2–1 lists the Student *t*-factor for 95% and 99% confidence limits as a function of n, the number of analyses run.

Note that the t values are very large for $n = 2$, but drop very precipitously at first and much more slowly for $n > 10$. Many analytical chemists will use twice the calculated standard deviation as the 95% C.L. and 3 times as the 99% C.L., often with very sparse data. As the table shows, this is strictly correct only if n is quite large, in the order of 50 for 95% C. L. and 15 for 99% C.L.

A chemist is often interested in the confidence which can be placed in the *mean* or average of n determinations. As mentioned previously,

Table 2–1 Student *t*-Factor

n	95% C.L.	99% C.L.
2	12.71	63.66
5	2.45	3.71
10	2.20	3.10
20	2.08	2.83
50	2.01	2.68
100	1.98	2.62
∞	1.96	2.58

we instinctively feel we can place more confidence in the mean than in the result of a single determination. This is correct, and statistics tells us that the standard deviation of the mean of *n* determinations (often confusingly called the "standard error of the mean" by statisticians) is given by:

$$s_{\bar{X}} = s/n^{1/2} \tag{2-5}$$

where $s_{\bar{X}}$ is the standard deviation of the mean, and *s* is the standard deviation association with a single determination. Since *n* is always greater than one, $s_{\bar{X}}$ is always less than *s*, indicating greater precision or smaller confidence limits for the mean value.

Equation (2–5) has interesting practical aspects due to the fact that $s_{\bar{X}}$ is inversely proportional to the *square root* of the number of determinations. Many chemists instinctively feel that the average of duplicate determinations should be twice as good as a single determination. This is a good example of a case where intuition is wrong. Equation (2–5) states that the imprecision of the mean of duplicates is $\frac{1}{2}^{1/2}$ or 0.707 times the imprecision of a single determination, or an improvement of only 30%. To obtain an estimation which is twice as good as a single determination, four determinations must be run, and nine must be run to obtain a threefold improvement in precision. It is apparent that the work required to effect a marginal improvement in precision soon becomes impractical, and a better method (with a smaller *s*) should be sought rather than continuing to run multiple analyses.

The standard deviation, as described above, is a number calculated from a set of *n* analyses applied to a single homogeneous sample. The standard deviation may also be conceptually attached to an analytical method, in which case it is often loosely identified with the "precision" of the method, where precision is defined as the reproducibility of the method. While the standard deviation may be a measure of the precision of the method, it should be noted that it is an inverse relationship, i.e., the smaller the standard deviation, the better the precision. In recent publications the author has noted the term *imprecision* associated with the standard deviation, a change which is surely for the better in improved communications.

When applied to an analytical method it is often found that the standard deviation is a function of the mean, i.e., of the level of analyte being determined. The function is often linear, in which case the relative standard deviation (RSD) is often found to be constant over the range of analyte concentrations encompassed by the method. The relative standard deviation is simply the standard deviation expressed as a percentage of the mean:

$$\text{RSD} = 100(s/\overline{X}) \tag{2-6}$$

In applications of the standard deviation as a measure of imprecision of a method, one should be aware of the fact that in some cases the standard deviation is independent of the analyte concentration and in others the relative standard deviation is constant. If the standard deviation is determined by multiple analyses of a "check standard" or standard reference material, it often may not be a true measure of imprecision of a result obtained at a very different concentration. Of interest in this regard is the data published by workers at the U.S. Food and Drug Administration[1] who showed that for a wide variety of analytes, and different instrumental methods of analysis, the relative standard deviation appeared to double for each tenfold decrease in analyte concentration. There does not appear to be a theoretical explanation for this empirically determined fact, but it is interesting to speculate on whether there is some undetermined natural law in operation.

PRACTICAL APPLICATIONS OF THE VARIANCE

The standard deviation is preferred as a measure of the dispersion of data for two reasons. The first is that units of expression of the standard deviation are the same as the mean, which makes for easy interpretation. However, there are other measures of dispersion which satisfy this criterion, e.g., the range, and the average deviation, which is simply the average of all the deviations, expressed as absolute values, i.e., without regard to sign. The reason the standard deviation is favored over these measures is because of its simple relationship to the variance, which is a measure of dispersion much more amenable to mathematical manipulation than the standard deviation. As described previously, the variance is simply the square of the standard deviation:

$$v = s^2 \tag{2-7}$$

The reason mathematicians prefer the variance as a measure of dispersion is that it can be shown that if the dispersion of a series of measurements is due to two or more causes, the overall variance is simply the sum of the variances due to the individual sources of variation. The only limitation on this statement is that the causes of variation must be independent of each other, and must individually be describable by the normal distribution.

A simple example will illustrate the usefulness of this relationship. Suppose we have a large batch of material and perform replicate analyses on it for some constituent. Suppose also that we suspect the material is not homogeneous so that the dispersion of the results is due both to inhomogeneity of the sample and imprecision in the analytical method. Then the overall variance is the sum of these two sources of variation:

$$v = v_b + v_a = s_b^2 + s_a^2 = s^2 \tag{2–8}$$

where s_a and v_a are the standard deviation and variance of the method, and s_b and v_b refer to the inhomogeneity. If s_a is known through analyses of other *homogeneous* materials, the dispersion due to inhomogeneity is easily calculated as:

$$s_b = (s^2 - s_a^2)^{1/2} \tag{2–9}$$

Mathematicians have evolved a whole sub-discipline of statistics called the Analysis of Variance (ANOVA) devoted to methods of designing experiments and evaluating data to determine the causes of variations in measurements.

A common problem in scientific measurement is to determine the significance of the difference in two measures of imprecision. For example, let us say we are comparing two analytical methods. We obtain a homogeneous batch of material and run n_1 determinations by Method 1 and n_2 determinations by Method 2. We calculate the standard deviations s_1 and s_2 and find that they are different, but not sufficiently so that we can decide unequivocally that the method with the smaller deviation is superior.

It is possible to solve this problem, with an arbitrary degree of confidence, by calculating the "*F*-ratio" and using statistical tables given in most textbooks on statistics. The *F*-ratio is simply the ratio of the larger of the two variances to the smaller, or:

$$F = s_1^2/s_2^2 \text{ where } s_1 > s_2 \qquad (2\text{–}10)$$

In the *F*-tables, one will find a separate table for each confidence level, e.g., 95%, 99%, 99.9%, etc. Choosing the desired table, one must enter, using $n_1 - 1$, and $n_2 - 1$, or the "degrees of freedom" used in calculating s_1 and s_2, to find the *F*-value. If the calculated *F*-ratio is higher than that in the table, one can say that the difference in imprecision is significant, with the degree of confidence represented by the chosen table.

This technique can be used in other ways. For example, one can compare results obtained by two different analysts analyzing the same samples, using the same method. Needless to say, this can be a powerful management tool in evaluating analyst performance.

Another useful application of the variance is in the technique called "pooling of variances." Suppose a number of samples of varying analyte concentration are analyzed in replicate. A variance and standard deviation can be calculated for each sample; but because the mean values are different, an overall standard deviation cannot be calculated using the simple formula of equation (2–2). It is possible to calculate an overall variance by summing the individual variances and dividing by the total number of samples run. In other words:

$$s^2 = [\Sigma s_i^2]/N \qquad (2\text{–}11)$$

where s_i refers to the standard deviation of the ith sample and N is the total number of samples (*not* the total number of analyses run).

The usefulness of equation (2–11) is most apparent in the case of duplicate analyses. If N samples are run in duplicate, equation (2–11) reduces to:

$$s^2 = [\Sigma R^2]/2N \qquad (2\text{–}12)$$

where R is the difference between the duplicate results or ranges, and N is the number of samples run in duplicate. It should be remembered that s calculated in this way is the standard deviation associated with a *single* determination, not the mean of duplicates, and that the degrees of freedom involved in s is N, not $2N$. Equation (2–12) then gives the laboratory supervisor a means of evaluating the imprecision of a method by simply splitting samples as they arrive at the laboratory and accumulating the results. Remember, the samples need not have the same analyte concentration. However, if the analyte concentration varies widely and the standard deviation is proportional to analyte concentration, the calculated standard deviation will be larger than the imprecision associated with measurement at a given analyte level.

To determine if the standard deviation varies with analyte concentration, the following test may be applied. Divide the set of duplicate results in two groups, those with high analyte levels and those with low. Calculate the standard deviation for each of the groups by equation (2–12). Now compare the two variances by the F-test to determine the significance of any difference in variance.

Another useful application of the variance is in the comparison of two means to determine if they are significantly different. A practical example would be in comparing two analytical methods to determine the extent of agreement between calculated results. If the same sample is run by each method, n_1 times by Method 1 and n_2 times by Method 2, two means $\bar{X}_1$ and $\bar{X}_2$ may be calculated, and in general will not be equal to each other. The problem is to determine if the difference is due to random variation or is real. The difference of the means is calculated:

$$d = \bar{X}_1 - \bar{X}_2 \tag{2–13}$$

This difference has two independent sources of variation, namely $s_{\bar{X}1}$ due to Method 1 and $s_{\bar{X}2}$ due to Method 2. The variance of the difference is then:

$$s_d^2 = s_1^2/n_1 + s_2^2/n_2 \tag{2–14}$$

as calculated from equations (2–8) and (2–5). The standard deviation may then be multiplied by the t factor corresponding to $n_1 - 1 + n_2 - 1$, or $n_1 + n_2 - 2$ degrees of freedom, and the desired confidence level. The result will give a range d ± C.L., Where C. L. is the confidence level. If the quantity zero falls within this range, it can be stated that there is no significant difference between the two methods at the chosen level of confidence.

REFERENCES

1. Horwitz, W., Kamps, L. R. and Boyer, K. W., Quality Assurance in the Analysis of Foods for Trace Constituents, *Journal of the Association of Official Analytical Chemists,* 63: 1344–1354 (1980).

3
STATISTICAL QUALITY CONTROL TECHNIQUES

QUALITY CONTROL CHARTS

In many manufacturing operations, chemical analysis is used to measure and control the quality of the product. Samples of product are taken, on a batch-to-batch basis or at appropriate times in a continuous process, and analyzed for some critical component or components. The results of the analysis are often plotted on charts which are used to demonstrate that product quality is being held within acceptable limits of variation.

The use of this type of Quality Control or QC charts was pioneered by W. A. Shewhart[1] in the first third of this century, not only for analytical chemical results, but for any measurement which could be related to product quality. Shewhart was concerned chiefly with the variation caused by the manufacturing process and less concerned with measurement variability. The latter was kept to a minimum by using average values of grouped measurements as points on the plot.

During the past twenty years, the concept of using Shewhart-type QC charts to control the quality of the measurement process itself has gradually taken hold in the analytical chemistry laboratory community. Rather than focusing on the variability of the samples with an assumed precise method of analysis, the focus is shifted to the variability of the analysis with an assumed homogeneous, stable matrix for analysis.

$\overline{X}$-Quality Control Charts

The process of establishing a QC chart is best described by an example. Suppose that a laboratory is analyzing multiple samples of a material on a regular basis using a well-established analytical method. Establishing a QC chart for this method has as its major objective the establishment of a process for determining whether the analytical system is in "control", i.e., that no sources of bias, or assignable error, have been introduced at any given point of time. Ancillary benefits will be the determination of the imprecision (standard deviation) of the method and the accuracy of the method in certain cases.

The first requirement for the establishment of a Shewhart $\overline{X}$-chart will be to obtain a large quantity of a stable, homogeneous "check standard," i.e., a material of the same substrate or type of substrate the method was designed to analyze. Obtaining a suitable material is not always easy. Standard Reference Materials (SRM's) for some analyses are available from the National Bureau of Standards, but there are many analyses for which they are not available. In addition, they are usually quite expensive and thus not suitable for routine use. The required stability may dictate storage at low temperatures, or under inert gas or controlled humidity conditions. Liquid matrices can usually be prepared with known concentrations of analyte, which enables the analyst to check accuracy of the method as well as precision, but solid matrices may not be as easily spiked and mixed to ensure homogeneity.

Once a suitable check standard has been obtained, a schedule is set up for running standards along with samples. It is generally recommended that one standard be run with each batch of samples, or, if large batches of samples are run simultaneously, one standard for every 10–20 samples. Needless to say, the check standard should be treated exactly like the samples in every respect.

When 15–20 analyses of the check standard have been accumulated, the $\overline{X}$-chart may be established as shown in Figure 3–1. The mean, $\overline{X}$, and standard deviation s, are calculated. The vertical axis of the chart represents the concentration of analyte, with a solid horizontal line drawn at $\overline{X}$. The horizontal axis simply identifies the check standard samples sequentially, i.e., number 1 is the first standard run, number 2 the second, etc.

Next, horizontal lines are drawn representing the 95% and, sometimes, the 99% confidence limits. For practical purposes, and assuming 15–20 data points have been used to calculate s, the values of $\pm 2s$, and $\pm 3s$ can be used for the confidence limit lines. The lines corresponding to $\pm 3s$ are often called the upper and lower "control

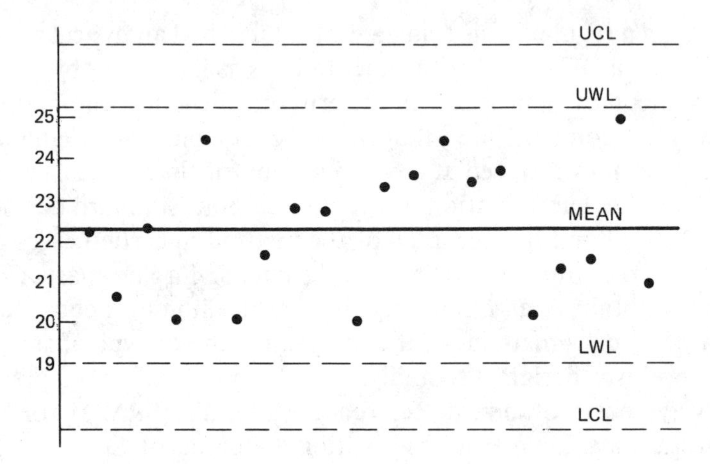

Figure 3-1 Typical $\overline{X}$-QC Chart

limits," or UCL and LCL, while the lines for the 95% confidence limits are similarly called the upper and lower "warning limits" or UWL and LWL.

Once the chart has been established, subsequent results obtained on the check standard may be plotted as they are generated. It is good practice to periodically (i.e., weekly or monthly) calculate the mean and standard deviation for the period chosen, as well as the cumulative mean and standard deviation. The current mean and standard deviation can then be compared to the cumulative values to determine if any differences are significant, as follows.

Assume N samples have been run to arrive at the cumulative mean $\overline{X}_c$ and standard deviation s_c. During the period under examination, n more samples have been run, yielding a mean of $\overline{X}$ and s. If the new standard deviation s appears to be significantly different from s_c, application of the F-test (see Chapter 2, page 15) with $N - 1$ and $n - 1$ degrees of freedom, will establish the significance to any desired level of confidence. If the new mean $\overline{X}$ appears to be significantly different from $\overline{X}_c$, the extent of significance can be estimated by using the method illustrated in Chapter 2 [(see equation (2–14)].

If the means and standard deviations are not significantly different, a new cumulative mean and standard deviation may be calculated. To do this, it is not necessary to go back to the raw data. For example, let $\overline{X}_t$ be the average for the total number of samples run ($N + n$). Then:

$$\overline{X}_t = (\Sigma X_i)/(N + n) = (N\overline{X}_c + n\overline{X})/(N + n) \qquad (3\text{–}1)$$

or the weighted average of the two means. Similarly, it can be shown that the new cumulative standard deviation s_t can be calculated from:

$$s_t^2 = [(N - 1)s_c^2 + (n - 1)s^2 + N\overline{X}_c^2 + n\overline{X}^2 - (N + n)\overline{X}_c^2]/(N + n - 1) \qquad (3\text{–}2)$$

all terms of which have previously calculated.

Practical Interpretation of $\overline{X}$-Quality Control Charts

Analysts will sometimes question the use of control charts on the basis that, if the mean and control limits are known, one can easily determine when the analytical system is out of control. While this is true, experience shows that the value of the control chart lies in detecting trends which may lead the system out of control if not corrected. The reason for this is that the human eye is a sensitive instrument for detecting *patterns,* and deviations from strictly random variation in a control chart become readily apparent. In the chart shown in Figure 3–2, for instance, no individual point is out of control, but the trend is clearly headed toward low results since seven of the last eight points are below the mean.

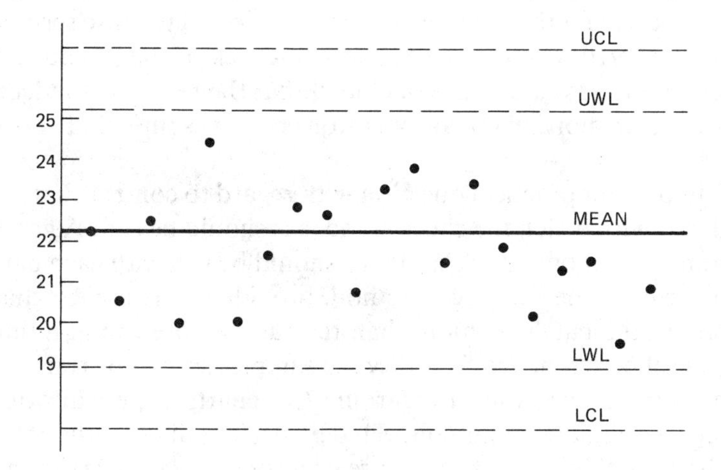

Figure 3–2 $\overline{X}$-Chart showing drift

To quantify this, probability theory tells us that the probability of any given point being above or below the average is clearly 50%, or ½, providing that the causes of variation are truly random. The probability of two consecutive points being on the same side of the line is 25% or ¼. Extending this, the probability of n points being on the same side of the line is $\frac{1}{2}^n$. If $n = 5$, the probability is $\frac{1}{32}$ or about 3% that this could happen through pure chance. The human eye can judge this at a glance if the data are charted, while it takes some concentration to pick this out of a table of data. If five or more consecutive points are on the same side of the average line, some action should be taken to detect sources of bias. New reagents may be prepared, instruments recalibrated, or perhaps new standards prepared.

With regard to the warning limits or 95% confidence limits, it should be remembered that one analysis out of twenty will exceed these limits, if the source of variance is random and normally distributed. Thus, a single analysis exceeding these limits should not be cause for alarm. In fact, a rough check of whether the limits have been properly calculated is to calculate the frequency with which these limits are exceeded. If the frequency is much lower or higher than 5%, the warning limits have been miscalculated, or an improvement or deterioration of precision has occurred.

On the other hand, the control limits, set at three times the standard deviation, have a probability of being exceeded by chance of about 0.3%. Any single result falling outside these limits should be immediately suspect, and the analytical system subjected to close scrutiny. If no cause of error can be found, and a re-check of the standard yields acceptable results, it can be concluded that the result was indeed due to chance, or more likely an operator error of some kind was committed.

An important practical question with regard to control charts is the frequency with which the check standards should be run. A good rule of thumb is that one check standard should be run with each batch of samples run by the analytical method, provided there are less than ten samples in the batch. If more than ten samples are run at a time, at least one check standard for every ten samples should be run.

One further observation regarding QC charts. One will often see such charts with the points connected by straight lines. Although there is no harm in this, there is also no reason for doing it since the chart in no way represents a continuous function, i.e., it is not rational to expect that if a standard had been run at a time between two points on the chart, the result would have fallen between the two values obtained.

$\overline{R}$-Quality Control Charts

For many analyses run in the average analytical chemistry laboratory, suitable materials to serve as check standards are not available or may be impractical to maintain because of cost, instability of analyte or matrix, or other reasons. Lacking check standards, the analyst must rely on other quality control operations, e.g., frequent instrument calibration, preparation of fresh reagent solutions, etc. to assure that results are as free of bias as possible.

It is possible, however, to measure and maintain control of the imprecision component of accuracy through the technique of periodically splitting samples and running duplicates. As described in Chapter 2, the standard deviation for a single determination can be easily calculated from a set of duplicate measurement on samples with widely different analyte concentrations.

To set up an $\overline{R}$-Quality Control chart, a series (15–20) of samples should be run in duplicate, and the range or absolute value of the difference in the duplicates calculated. More than two replicates of each sample can be run, of course, and a range calculated, but common practice and cost-effectiveness generally limit the number of replicates to two.

An average value for the range can then be calculated and a chart prepared in which the ranges are plotted consecutively as they are run in a manner similar to the $\overline{X}$-chart. Figure 3–3 is an example of an $\overline{R}$-chart.

Distribution of Ranges

The ranges, or differences between duplicates, will display a somewhat different distribution than that of the individual results about a mean. Youden[2] has presented a table for the distribution of differences between duplicates (see Table 3–1).

Based on this distribution, it can be shown that 50% of the ranges should be above the line corresponding to $0.845\overline{R}$, $2.456\overline{R}$ corresponds to the 95% limit, and $3.27\overline{R}$ corresponds to 99% limits.

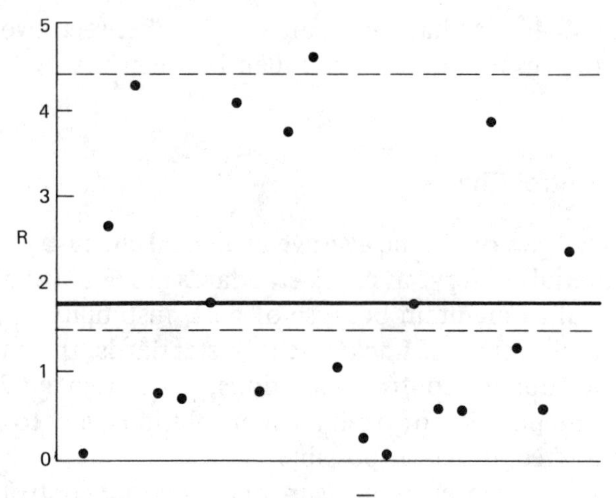

Figure 3–3 Typical $\overline{R}$-QC Chart

Therefore, lines corresponding to these limits should be drawn on the $\overline{R}$-QC chart.

Interpretation of the $\overline{R}$-QC Chart

Interpretation of the $\overline{R}$-chart is similar to the $\overline{X}$-chart. If 5 or more consecutive points are plotted and are above the 50% line, it may be concluded that the analysis is tending toward being out of control, and corrective action should be taken. One point above the 95% line for every 20 duplicates is to be expected on the average, but a point above the 99% line should be viewed with extreme suspicion. A value of zero is, of course, a perfectly legitimate value for the range, but if too many

Table 3–1

DIFFERENCE	%
Less than $\overline{R}$	57.5
Between $\overline{R}$ and $1.5\overline{R}$	19.4
Between $1.5\overline{R}$ and $2.0\overline{R}$	12.1
Between $2.0\overline{R}$ and $2.5\overline{R}$	6.4
Greater than $2.5\overline{R}$	4.6
	100.0

Source: The Association of Official Analytical Chemists. Reprinted by permission of the publisher.

zeroes are found, it is likely that the results are not being carried out to the correct number of significant figures.

As has been pointed out previously, the range values can be used to calculate the standard deviation of the analysis, using equation (2–12). As in the case of the $\overline{X}$-chart, the standard deviation should be calculated and compared to the cumulative standard deviation. The significance of any difference in the two standard deviations can be determined on the basis of the F-test.

After calculating the standard deviation for the new month or week, the total standard deviation may be calculated from these data and the cumulative data. It is not necessary to recalculate the new cumulative statistics by going back to the raw data. Assume a cumulative $\overline{R}_c$ and s_c have been previously calculated from N pairs of duplicate analyses and the new monthly or weekly statistics are $\overline{R}$ and s, based on n pairs. The new cumulative statistics $\overline{R}_t$ and s_t can be calculated from:

$$\overline{R}_t = (N\overline{R}_c + n\overline{R})/(N + n) \qquad (3\text{–}3)$$

and

$$s_t = [(Ns_c^2 + ns^2)/(N + n)]^{1/2} \qquad (3\text{–}4)$$

The frequency with which duplicates should be run is similar to that of the $\overline{X}$-QC charts, i.e., at least one duplicate for every ten samples, or with each batch of samples, if there are less than ten samples in a batch.

The major disadvantage of the $\overline{R}$-chart is that it tells the analyst nothing about the possibility of bias or constant error present in his results. Since the difference between analyses is calculated, error due to bias is effectively cancelled, leaving only the error due to random variation. Another disadvantage is that the technique cannot be used if the standard deviation, and hence the range, is related to the level of analyte concentration. As discussed previously, this can be determined by dividing the set of duplicates into those at high concentration levels and those at low, and calculating separate standard deviations for the two groups. Application of the F-test will determine the significance of any difference between the two groups. Another method is to plot range versus analyte concentration and use visual examination to determine if there is any correlation between the two. A refinement would be to calculate the correlation coefficient, but this is seldom necessary in a practical case.

A practical advantage of the $\overline{R}$-chart is that no special check standard sample is required. Thus, it can be used with analyses where check standard materials are difficult or impossible to obtain, e.g., the "suspended solids" analysis on waste water. In some analyses, the analyte is inherently unstable, and check standards would slowly degrade with time, an example being the "xanthate sulfur" test on viscose solutions.

In the application of $\overline{R}$-charts, it is simply necessary to split one sample in each batch in two and analyze each half as though it were a separate sample. If the supervisor can do this without the knowledge of his analysts, the resulting data will give an excellent picture of the true run-of-the-mill precision in his laboratory since the analyst will not know he is running duplicates and hence will be less likely to take special care with these samples, as he may if he is conscious of running a "standard sample." Another advantage is that the data are based on actual samples and hence reflect all of the variations which can occur due to the substrate, e.g., variations in homogeneity or the actual physical state (adsorbed, free, crystal size, etc.) of the analyte, or interferences which may be present.

Spiked Sample Control Charts

In some analyses where check standard materials are not easily found or prepared, the technique of using "spiked" samples is used for quality control. In this technique, a known amount of analyte is added to a sample and analyses are run before and after addition to the spike. The difference in the two analyses is then used to calculate the percent recovery of the known amount of analyte. $\overline{X}$-charts are prepared and used in the usual manner except that percent recovery is plotted instead of the mean or concentration of the spike.

While the use of spiked samples is valuable to some degree, the technique is not as useful as the check standard or duplicate sample methods, and the analyst should be aware of the dangers in interpreting results. From a statistical viewpoint, the calculated standard deviation will always be greater than the true standard deviation because two determinations are used in the calculations. If s is the observed standard deviation, s_o the standard deviation of the determination of the sample, and s_p that of the spiked sample, equation (2–8) states that:

$$s^2 = s_o^2 + s_p^2 \qquad (3\text{–}5)$$

If $s_o = s_p$ (not always true), then:

$$s = 2^{1/2} s_o = 1.404 s_o \qquad (3\text{–}6)$$

If the standard deviation is a function of the analyte concentration, s_p will generally be larger than s_o and the observed standard deviation will be even greater.

Another disadvantage of the spiked sample method will be illustrated by an example. In the determination of pesticide residues on materials such as agricultural products, soils, etc. there is often considerable sample preparation prior to the final measurement by gas chromatography. The sample may be ground, mixed, extracted with a solvent, the extract "cleaned up" by passage through a chromatographic column, and then concentrated by evaporation, prior to injection into the GC. The spiking is often done at the last step, i.e., a second aliquot of the extract is taken and spiked just prior to injection. The result, of course, is completely meaningless as far as interpretation of the precision and accuracy of the overall method, since the statistics apply only to the spiking, injection, and measurement steps. Any variation which occurs in sample preparation is excluded from the measurement.

The original sample, of course, should be spiked. However, this in itself can be a cause of problems, especially if the sample is a solid material. Unless the whole sample is used for spiking, problems in mixing and homogeneity can occur. Furthermore, there is no guarantee that the spiked material is identical to the analyte in the substrate. It may be in a different physical state, for example less tenaciously adsorbed on the substrate, or of different crystal size or habit which could affect its solubility, or in a different chemical state due to reaction of the analyte with components of the substrate.

For these reasons the use of spiked samples is not recommended for QC charts unless the check standard or duplicate analysis method cannot be applied.

One exception could be in the case where an investigation into the *causes* of variation may be carried out. A set of duplicates may be analyzed to obtain the overall variance of the method, and a set of spiked samples used to measure the variance of the measurement process, as in the GC analysis previously described. The difference between the two variances then gives the variance due to the sample preparation step.

Cusum Quality Control Charts

One technique for control charting which seems to be somewhat in vogue these days is the use of "cusum" charts. The word *cusum* is a contraction of "cumulative sum," because what is plotted is not the actual value of the result obtained on a standard or the range of duplicates, but the cumulative sum of the differences between the result and the previously established mean, or known value of the standard. In other words, the difference between each result and the mean is calculated, and the second difference added to the first, the third difference is added to the sum of the first and second, etc.

Obviously, if all the differences are positive or all are negative, the plotted results show a line with a positive or negative slope indicating that the system is out of control. If the differences are randomly positive or negative, the plotted line will be more or less horizontal. Various techniques are available for diagnosing the plots.

The cusum plot is widely used in industrial quality control of manufactured products because it is more sensitive to trends in data than $\overline{X}$ and $\overline{R}$ charts. However, the author prefers to use $\overline{X}$ and $\overline{R}$ charts in analytical chemistry QC because of the fact that they are more easily interpreted (often by a glance) and lead to easy calculation of useful statistics which is not the case with cusum charts without additional calculations.

BLIND SAMPLES

One of the most valuable techniques for evaluating the quality of the entire analytical system, i.e., personnel, instruments, methods, etc., is the use of blind samples. A blind sample is one which is of known concentration, but is submitted as a routine sample and without the analysts being aware of the fact that there is anything special about the sample. Usually only the supervisor or Quality Assurance Director knows the nature of the sample.

The best blind sample technique uses a check standard of the same type as described under the $\overline{X}$-QC chart system. This material can be submitted blind on a periodic basis, and the results then plotted in the usual way, by the supervisor or QA Director. An alternative is to submit duplicates on a blind basis, using an $\overline{R}$-chart to estimate the imprecision of the method.

The advantage of a blind sample QC system is that it evaluates the true quality of the system, since the known sample is treated exactly the same as the run-of-the-mill samples regularly analyzed. Invariably it will be found that the imprecision found in the blind sample system is greater than that determined on the basis of the $\overline{X}$-chart. The reason for this is that when an analyst analyzes a sample known to be a check standard he will, consciously or subconsciously, tend to be more careful than with the routine sample. The temptation is always present to re-run the sample if a result is obtained which is not perceived as "good" by the analyst, and unfortunately the temptation is not always resisted. The result is that the standard deviation on check samples is always artificially small compared to the standard deviation based on a blind sample system.

Although the blind sample system is undoubtedly superior, there are very real practical difficulties involved in setting it up. It is necessary, for example, to use a dummy client name for the origin of the sample. If a real client's name is used, there is danger of the result being reported to the client instead of being intercepted by the supervisor or QA Director, with resultant confusion. On the other hand, if a fictitious name is used, there is the danger of the analyst discovering that the client is non-existent and that the sample is therefore a blind standard. One way of extending a blind sample system's integrity is to submit the samples on an infrequent basis. For example, the blind standard may be submitted every two weeks. After a year, 26 determinations will have been run which is an adequate number for calculations of accuracy and imprecision.

PROFICIENCY OR PERFORMANCE EVALUATION SAMPLES

These are samples which are distributed to a laboratory for analysis by an organization external to the laboratory, e.g., a government agency, accrediting body, or a client. They are often submitted in support of a system for accrediting or certifying a laboratory's competence. For example, the U.S. Environmental Protection Agency submits performance evaluation samples to laboratories certified by EPA or the states to analyze potable water. The National Institute of Occupational Safety and Health (NIOSH) distributes samples (PAT or Performance Analytical Test) samples on behalf of the American Industrial Hygiene Association as part of their accreditation process.

In other cases, they are submitted by professional organizations whose members are interested in the competence of their laboratories. Examples are the check feed program of the American Association of Feed Control Operators (AAFCO) or the Smalley check sample program of the American Oil Chemists Society (AOCS). In these latter types of programs, participation is strictly voluntary, while in those concerned with accreditation or certification, participation is mandatory.

In those programs involving accreditation, the samples submitted contain analytes at known concentrations, which are unknown to the laboratory. The concentrations are known, either because they were prepared that way, or because they were exhaustively analyzed by the laboratories of the organization or by laboratories in which the organization has confidence, or both. The laboratory being evaluated is sent a report after the data are received, comparing its results to the known values. Accreditation is either affirmed or denied based on whether the laboratory's results are within "acceptable" limits or not. The limits of acceptability are usually either the 95% or 99% confidence limits. If the 95% limits are used and a laboratory exceeds them, accreditation is usually provisional, pending analysis of another proficiency sample, while if 99% limits are used, accreditation is usually denied until the laboratory successfully analyzes the second sample.

Maintaining or achieving accreditation or certification of a laboratory is important from a professional, economic, and legal standpoint. Therefore, there is great pressure on a laboratory's staff to achieve acceptable results on proficiency or performance evaluation samples. Multiple analyses are run if sufficient sample is available, with only averages being reported; instruments are calibrated before and after analysis; standards are run before and after; fresh reagents are prepared; and the best analyst assigned to the task. The pressure may be great enough to tempt cheating, e.g., running the concentrate supplied instead of following dilution instructions, because the precision may be better on the concentrated sample. Instances have been known where laboratories have sent samples to other laboratories for comparison purposes.

This is not to say that proficiency sample testing is a waste of time. Undoubtedly, the process is a valuable one for the accrediting organization, since it will weed out those laboratories which are so bad they should not be operating, and will force marginal laboratories to upgrade their operations to an acceptable level.

However, laboratory management and supervision should be very wary of using the results of proficiency sample testing as representative of the quality of the data generated in their laboratory. If all samples were analyzed with the same care as proficiency samples, this would be true, but the cost of analysis for routine samples would undoubtedly be unacceptable. Clients or users of laboratory services should be especially aware that results on proficiency samples are not necessarily representative of the overall quality of the laboratory's work.

The other type of proficiency testing mentioned previously, that involving voluntary participation, is much more useful to the laboratory supervisor. Since there is no pressure for accreditation or certification, these samples may be submitted blind by the supervisor or QA Director. The laboratory generally receives a full report, with statistical analysis, including the data from all participating laboratories, suitably coded for confidentiality. The laboratories are often ranked, according to their closeness to the "true" value or the average of all participating laboratories. Such ranking should be viewed with caution, because of the random variation in all analyses; but if a given laboratory consistently is higher or lower than average, it may well be indicative of the laboratory's overall quality in comparison with other laboratories.

REFERENCES

1. Shewhart, W. A., *Economic Control of Manufactured Products,* New York: Van Nostrand Co., 1931.
2. Youden, W. J. and Steiner, E. H., *Statistical Manual of the Association of Official Analytical Chemists,* Washington, D.C.: Association of Official Analytical Chemists, 1975.

4
ANALYTICAL METHODS

INTRODUCTION

It is almost a truism to state that one of the major sources of confusion in the world is poor communication. Communication between human beings requires at least three components: the person transmitting information, the person receiving information, and the medium of transmission. In the analytical laboratory, the medium by which management informs the workers in the laboratory of the procedures by which analyses are carried out, is the analytical method. It is one objective of quality assurance to ensure that the workers are unambiguously informed and that unintentional variations from the desired procedures are kept to a minimum.

In an analytical laboratory, all analytical methods should be in written form. Although this seems obvious, it has been the author's experience that in many laboratories today, this is not the case, or if written methods are used, they are in very sketchy form. In many laboratories, new workers are simply trained by experienced workers in the procedures used, with little reference to a written procedure. In many labs, procedures are simply outlined on index cards, written on scraps of paper, or pencilled in margins of textbooks or notebooks.

The hazards in this type of operation are obvious. If methods are not explicitly written out and available to the analysts, the possibility of intentional or unintentional variations in the method is always present. Non-degreed personnel in particular, since they usually do not have the scientific background to completely understand the reasons for various operations, may be tempted to take shortcuts, eliminate steps, substitute reagents, etc., sometimes in a commendable desire to simply increase productivity. Such "creative technicianship" usually results in loss of accuracy, precision, or both.

On the other hand, unless the method is written clearly and unambiguously, even professional chemists may misinterpret the instructions, leading to poor results. This is undoubtedly one of the major reasons for the loss of precision when a method is tested "round-robin" for interlaboratory precision.

Thus, a good quality assurance program requires that methods be written out, that they be written in considerable detail, and that personnel be informed that deviations from the written method are not allowed without express permission of the laboratory supervisor.

CHOOSING METHODS

The proper choice of an analytical method is subject to many considerations. The method chosen must first of all be consistent with the level of analyte being measured and must also be appropriate for the particular substrate or matrix containing the analyte. Other factors to be considered are: the availability of instruments and other equipment, the speed with which results are required, cost, convenience, safety factors, and, of course, the accuracy and precision of the method. Some of these factors, e.g., speed, cost, and accuracy are dictated by the client's needs, while others are dictated by the practical consideration of the operating laboratory

In general, the prudent laboratory supervisor will choose methods which are available in the literature, provided such methods can be found and that they are consistent with the factors indicated above. The development of an analytical method from scratch can be a costly proposition if done properly, and if not done properly can lead to problems and frustration in later applications. The temptation to simply devise a method based on one's knowledge of chemistry and then immediately put it into practice after a couple of samples have been run, should be strenuously resisted. On the other hand, simply abstracting a method from the general literature can also have its pitfalls.

Standard Methods

By far the best literature methods are those produced by various standard-setting organizations. The reason for this is that methods adopted by these organizations are thoroughly investigated and tested in numerous laboratories before being designated as "standard." In addition, most of these organizations are made up of individuals with a vested interest in producing test methods which are accurate, and practical, for the analysis of various commercial products. Therefore, the laboratory which uses methods produced by these organizations is less vulnerable to criticism concerning the use of proper methods, than if a method is used which does not have the "official" or "standard" designations of a nationally known organization. Some of the organizations producing such standards in the United States are:

ASTM - American Society for Testing and Materials
ANSI - American National Standards Institute
AOAC - Association of Official Analytical Chemists
CTFA - Cosmetics, Toiletries and Fragrances Association
USP - The United States Pharmacopeia
AATCC - The American Association of Textile Chemists and Colorists

There are literally hundreds of organizations in the United States producing standards, but only a fraction of these standards relate to chemical analysis. In addition, of course, there are numerous organizations in other countries producing standards or official methods. In general, a "standard" or "standard method" is one which has been agreed upon by members of a standard-setting organization, e.g., ASTM or ANSI, which operates on a consensus principle, i.e., all members (and sometimes non-members) who have an interest in the product, including producers, government officials, and consumers, are permitted to have input into the generation of the standard. An "official method," on the other hand, is usually produced by an organization, e.g., a government agency or professional society with a mandate of some sort to produce testing methods. "Standard" methods may become "official" methods by adoption by a government agency.

Official Methods

Government agencies which produce official methods are generally those involved in regulatory activities where lack of official testing methods could easily lead to chaos. Because of their importance, and careful scrutiny by the industries being regulated and by consumer groups, these methods are generally thoroughly validated before they are declared official. Examples of official government methods are those published by the Environmental Protection Agency for testing potable water, the National Institute for Occupational Safety and Health methods for sampling and analysis of airborne contaminants, and the Food and Drug Administration's methods for food testing.

Wherever possible and practicable standard or official methods should be adopted for routine laboratory analysis. These methods have been thoroughly tested, are widely recognized, and generally have associated with them a body of knowledge regarding accuracy, precision, interferences, etc. which requires no additional work in the laboratory. Because of the long time involved in establishing a standard or official method (sometimes years) they may appear to be somewhat outdated, but it is always better to be safe than sorry in choosing a method.

Literature Methods

The periodic literature on analytical chemistry is extensive. Not only are there general purpose journals, e.g., *Analytical Chemistry, Analytica Chimica Acta, Journal of the AOAC,* etc., but there are many specialized journals, e.g., those dealing in food chemistry, pharmaceuticals, paper, plastics, etc. which publish occasional articles containing or featuring analytical methods. The patent literature and house organs, especially of instrumentation manufacturers, are also fruitful sources of analytical methods.

Any analytical method abstracted from the literature, even from prestigious journals such as *Analytical Chemistry* should always be approached with caution. This is especially true if the application in the laboratory is somewhat different from that reported in the original article, e.g., an application to a different substrate or matrix, a somewhat different chemical or physical form of the analyte, or even using instruments of different capabilities. Even if the application is straightforward, caution is advised. The authors of the original article are usually far from objective or unbiased in their assessment of usefulness, convenience, accuracy, or precision of their brain child.

VALIDATION OF ANALYTICAL METHODS

Validation of New Methods

With the exception of standard or official methods as described previously, any new method or major modification of a method which is being considered for use in routine testing of large numbers of samples should be subjected to a validation study in the laboratory. The purpose of the validation study is to demonstrate that the method is appropriate for the analysis of the substrate to which it will be applied, and to gather preliminary information on precision and accuracy inherent in the method itself.

To this end the method should first be written out explicitly, and the validation study should be carried out by an experienced analyst. This helps avoid variability due to possible misinterpretation of the procedures involved and variability due to inexperience on the part of the analyst or the learning curve of a less experienced analyst.

In the next step a series of samples should be obtained or prepared which cover the range of analyte concentration normally expected to be encountered in the substrate. The ideal situation is when samples can be obtained of known concentration of analyte, but this is often not possible. Failing this ideal, samples can be constructed by spiking substrate with known amounts of analyte. The substrate used should preferably be free of analyte before spiking. Proper precautions should also be taken to insure homogeneity of the spiked sample and that the spiked analyte is in the same physical and chemical state as the analyte naturally occurring in the substrate.

Four or five samples should be obtained and then subjected to replicate analyses using the method exactly as written. Four or five replicates of each sample should be run, preferably in random order and on different days, to subject the method to whatever uncontrolled or unrecognized variables may be operable. As mentioned previously all analyses should be run by the same analyst who should be experienced in the type of operations involved in the method, since we are not interested in between-analyst variance, but only in those factors likely to affect the method *per se*.

When all analyses have been run the data can be evaluated as follows. First, the variance is calculated for each sample based on the replicates run on that sample. The variances for the two samples representing the highest and lowest analyte concentrations should then be compared by the F-test (Chapter 2, page 15) to determine if there is a significant difference between them at the 95% confidence level. If the difference is not significant the variances for all samples can be pooled to give the overall variance, and hence standard deviation, for the method, using the following equation:

$$s^2 = \Sigma n_i s_i^2 / (\Sigma n_i - N) \tag{4-1}$$

where s is the standard deviation of the method, n_i the number of replicates of the ith sample, s_i^2 the variance of the ith sample, and N is the total number of samples, i.e., four or five in the above examples.

If the F-test indicates significant variation of the imprecision with the level of analytes, a plot may be constructed of s versus $\bar{X}$, or better yet the 95% or 99% confidence limits based on s and the appropriate t-factor (of Chapter 2) versus $\bar{X}$. This can be used to interpolate s or the confidence limit for levels of analyte not covered in the study. On the other hand, in many analytical methods s is linearly proportional to $\bar{X}$, and the relative standard deviation remains constant over a considerable range. This can be tested by dividing s by $\bar{X}$ for each sample and examining the results to see if RSD is relatively constant. Further testing may be necessary to establish this point unambiguously.

In the next step a graph is constructed, plotting on the y-axis the mean of the various replicates, and on the x-axis the known, true values of the concentration. The plotted points should be compared with a straight line drawn through the origin with a slope of unity. If the points do not fall on or very near the line, the "least squares" straight line should be calculated and drawn on the graph. This is done as follows. Assume we have N pairs of values y, x, where the y's are the mean analysis values and the x's are the known concentrations. Calculate the following:

$$S = \Sigma x \qquad D = NQ - S^2$$
$$Y = \Sigma y \qquad Q = \Sigma x^2$$
$$P = \Sigma xy \qquad L = \Sigma y^2$$

Then the slope of the least squares line will be given by:

$$m = (NP - SY)/D \tag{4-2}$$

and the intercept by:

$$b = (QY - SP)/D \tag{4-3}$$

These two results are the parameters of the equation:

$$y = mx + b \tag{4-4}$$

which defines a line of slope m and intercept b, which is the "best" line that can be drawn through the points, in the sense that it minimizes the deviation of the points from the line.

The question then arises of, given a slope m different from unity and an intercept b different from zero, what is the probability that these represent merely a random variation from a true slope of unity and intercept of zero? If the above experiment were repeated there is high probability that the new slope and intercept would be different from the first one.

Mandel and Linnig[1] addressed this problem in a paper published more than 25 years ago. They showed how to construct a "confidence band" around the least squares straight line, corresponding to a given confidence level, e.g., 95% or 99%. If the theoretical line of unity slope and zero intercept lies within the confidence band defined by the experimental data, then it is possible to state with the desired level of confidence that the experimental slope and intercept are not significantly different from unity and zero respectively.

The confidence band of Mandel and Linnig is represented by two arms of a hyperbola, which may be calculated as follows. Calculate the following expressions:

$$s^2 = 1/(N-2)[L - Y^2/N - (D/N)m^2] \tag{4-5}$$

and

$$K = (2Fs^2)^{1/2} \tag{4-6}$$

where F is the F-factor determined from a table giving F at the desired confidence level, and entered with 2 and N–2 degrees of freedom (Chapter 2, page 15). The two arms of the hyperbola are given by the equation:

$$y = b + mx \pm K[1/N(1 + (x - \bar{x})^2/D/N^2)]^{1/2} \tag{4-7}$$

Although equation (4–7) seems somewhat formidable it is relatively easy to handle with a simple electronic calculator. From equation (4–7), calculate two values of y for a series of assumed x values, one value of y corresponding to the positive sign preceding K and the other corresponding to the negative sign. When plotted on the graph the positive sign values will chart the upper branch of the hyperbola and the negative values the lower branch. Figure 4–1 is a typical example.

It is now possible to examine the graph and determine if the theoretical line of unity slope and zero intercept falls within the confidence band established by the two branches of the hyperbola. If the line is well centered there can be little doubt that the method is unbiased. If the line is not well centered, but falls close to the limits, it may be wise to recalculate using a tighter confidence limit (say 99% rather than 95%).

Comparison of Analytical Methods

The previous paragraphs have dealt with the case where a completely new analytical method was being validated. A frequent situation which arises is the comparison of a new method with an old one. The new method may be based on a completely different scientific princi-

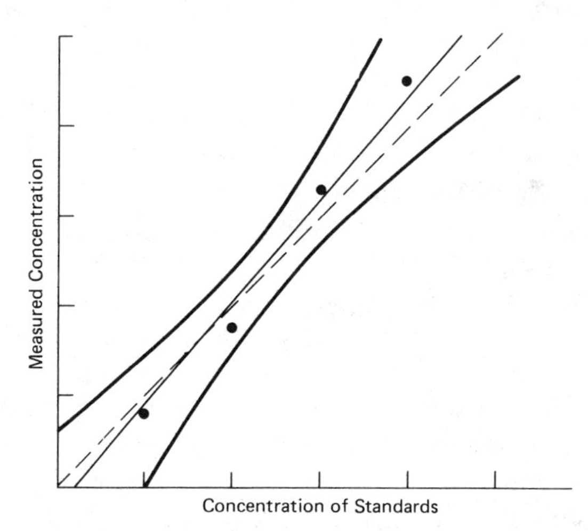

Figure 4–1 From: Mandel, J. and Linnig, F. J., *Anal. Chem 29,* 743–749 (1957).

ple, or may be a modification of the old method, e.g., the use of a larger sample size, or more concentrated or dilute, reagent solution.

The technique described above may be used here as well, except that it is not necessary to prepare samples of known concentration. A series of samples may be obtained which cover the normal range of analyte concentrations and which are then analyzed in replicate by both methods. It is assumed in this instance that the mean of the results obtained by the old method represents the "true" concentration of the analyte, and these are plotted on the x-axis, while results obtained by the new method are plotted on the y-axis. Standard deviations may be calculated for both methods and compared by the F-test to determine if any differences in imprecision are significant. Calculation of Mandel and Linnig's confidence band will determine whether results from the new method are at least as reliable as the old method with any desired degree of confidence.

Ruggedness Testing of Methods

An analytical method consists of a series of operations which result in the generation of a set of numbers which, when combined in a prescribed manner, yield a single number which we hope describes a property of the sample. Many variables may be involved in the operations, e.g., weighing, drying in an oven, filtering, addition of reagents, time of heating, temperature, etc. Some methods will be sensitive to minor variations in these variables, while others will not be. The latter type of method has been termed "rugged" by Youden[2], and he has described a method for evaluating the "ruggedness" of an analytical method. It is recommended that all new analytical methods which have not been thoroughly investigated, be subjected to a Youden "ruggedness" test before being adopted as an authorized method by an analytical laboratory.

The Youden ruggedness test is an efficient one in that it enables the analyst to evaluate the effect of seven different variables with only eight analyses. The first step is to determine the seven variables to be evaluated. These could be: sample size, drying time, drying temperature, reagent concentration, rate of stirring, type of filter, etc. Establish two levels for each variable, e.g., reagent concentration ± 10% of the recommended concentration, or heating time ± 10% of that recommended, or two different types of filter media. The two levels should be of the type or magnitude which might be encountered in normal use of the method. If seven variables cannot be identified, one or two dummy variables may be used. Assign seven capital letters, *A* through *G*, to one level of each variable and seven lower-case letters to the other level. For example, the letter *A* may be assigned to 1.2 gram sample size and the letter *a* to 0.8 gram sample size. The analyses are then carried out according to the scheme given in Table 4–1.

From the results *s* through *z* we can calculate the effect of each of the variables by dividing the eight determinations into two groups based on the letters. The effect of going from *A* to *a*, for example, is given by the difference between $(s + t + u + v)/4$ and $(w + x + y + z)/4$. Note that all of the other variables are present twice at both levels in each of these two groups. Thus, effects due to variables *B* through *G* are cancelled by *b* through *g*. Similarly, the effect of the seventh variable, i.e., changing *G* to *g*, is given by the differences between $(s + v + x + y)/4$ and $(t + u + w + z)/4$, and again the effects of the other six variables are cancelled.

Collect the seven average differences *A*–*a* through *G*–*g* and examine them. Any variable with a real effect will have a difference considerably larger than the others. Such variables should be given special attention in the final written version of the method, spelling out the fact that they must be controlled with greater than normal attention.

Incidentally, the standard deviation calculated from the eight results *s* through *z* is an excellent measure of the level of imprecision to

Table 4–1

	Combination or Analysis Number							
Factor Value	1	2	3	4	5	6	7	8
A or a	A	A	A	A	a	a	a	a
B or b	B	B	b	b	B	B	b	b
C or c	C	c	C	c	C	c	C	c
D or d	D	D	d	d	d	d	D	D
E or e	E	e	E	e	e	E	e	E
F or f	F	f	f	F	F	f	f	F
G or g	G	g	g	G	g	G	G	g
results	s	t	u	v	w	x	y	z

Source: The Association of Official Analytical Chemists. Reprinted by permission of the publisher.

be expected in routine use, since you have deliberately introduced the type of variations which might normally be expected to occur.

WRITING ANALYTICAL METHODS

As mentioned in the beginning of this chapter, all methods which are used for standard, repetitive testing in the laboratory should be in written form and kept in an area of the laboratory where they are readily accessible to all analysts. This is an important factor in maintaining a good quality assurance program. Furthermore, the method as written should correspond exactly to the method as it is actually used in the laboratory. Laboratory supervisors should monitor operations closely to be certain that deviations from the written methods are not occurring. Deviations are not necessarily bad and in fact may represent an improvement of one type or another. If so, and if it is established that the deviation does not have a deleterious effect on quality of data, then the method should be rewritten.

In writing an analytical method, two things should be kept in mind. The first is that it should be as clear as possible, with minimum ambiguity, so that a chemist or experienced technician unfamiliar with the method would be able to use it with no instruction from the supervisor. The second is that enough information should be included so that the analyst can interpret the results obtained. No professional chemist should use a method whose scientific basis is unclear.

Many different formats are used in writing analytical methods. The following is one which the author recommends on the basis of practical experience. Each part of the method is listed in the order in which it appears. No section should be left out. If the information required for that section is not available, that fact should be explicitly stated.

Format

1. Number of the Method – All methods authorized for use in the laboratory should have an assigned number for traceability purposes. The number can be noted in the analyst's notebook and should also appear on the report of results. The methods may be simply numbered consecutively or various schemes may be used to classify methods, e.g., use of initial letters to designate substrates (*F* for food, *W* for water, etc.). The important point is that only *one* number should be used for any given

method, and only *one* method should be assigned to each number. Although this seems obvious, experience shows it is not always followed. The temptation to give a blanket number to methods involving slightly different procedures is not always resisted.

2. Date Authorized.
3. Title – Care should be taken with the choice of title. The title should be brief and should contain the name of both analyte and substrate, e.g., "Determination of *Nitrogen* in *Animal Feeds.*" If more than one method exists for a given analyte/substrate combination, the name may include a designation of the measurement method, e.g., "*Spectrophotometric* Method for Determination of *Color* in *Wastewater.*" Occasionally, the method used will depend on the level of the analyte present and should be noted in the title, e.g., "Determination of *Low Level Phenol* in *Wastewater.*"
4. References – Include references to the literature on which the method is based, or in-house documentation of validation studies.
5. Scope – The scope should show the range of analyte concentrations for which the method is useful, the type and nature of the matrix to which it can be applied, and an estimate of the time required for a single analysis. It should also indicate interfering substances. In other words, it should enable the analyst to decide quickly whether the method is suitable for the particular analytical problem with which he may be faced.
6. Basic Principles – This section should describe the chemical, physical or biological principles on which the method is based. Unusual chemical reactions should be written out, separation processes described briefly, and the effect of interfering substances should be described. Methods used to eliminate interference should also be described.
7. Apparatus and Reagents – This section should describe instruments and unusual apparatus which are required. Common laboratory equipment, e.g., pH meters and analytical balances need not be listed unless special capabilities are required such as the need to measure 0.001 pH units or micro or semi-micro weighing capability. Similarly, common glassware need not be listed, but specialized pieces, e.g., a Soxhlet extractor should be mentioned.

Reagents should be fully described, including chemical name, purity and description of method of preparation for those which need to be prepared prior to analysis. Shelf life should be given where stability may be a problem.

8. Safety Precautions – Any safety precautions peculiar to the analysis should be described. These might include: necessity for working in a hood, steps to avoid hazardous reactions such as explosions, need for special safety devices or clothing, and special precautions to deal with hazardous waste disposal. The reasons for the safety precautions listed should be given, so that the analyst may more readily assess the degree of hazard represented.
9. Procedure – This is the heart of the method and should be written with special care. When writing the procedure, put yourself in the position of the reader, and imagine you are a recent B.S. chemistry graduate and have been given this method and asked to analyze a sample. The following guidelines may help in writing this section.
   (a) Follow a strict time sequence, exactly as the test is run. There is nothing more frustrating than to suddenly come upon an instruction such as: "Add 20 ml of a 1:10 dilution of a 50:50 mixture of reagents *A* and *B*, which have previously been mixed and filtered." This means the analyst must turn his attention from the sequence of operations in the procedure to mix two reagents, filter the mixture and dilute 1:10 before continuing the main sequence of the procedure.
   (b) Avoid abbreviations unless you are sure they are commonly understood. Remember that an abbreviation may be well understood by you and your current staff, but not familiar to a new employee who has just joined the staff. If convenient, an abbreviation may be defined the first time it is used and then used after that.
   (c) Be specific. Don't say "neutralize with HCl," when what you mean is "add 0.1M HCl dropwise to a pH of 7.0 ± 0.2." On the other hand, it is not necessary to spell out in detail common laboratory operations, e.g., weighing or titrating.
   (d) Indicate critical steps in the analysis and the consequences if care is not taken.

(e) Use short sentences and avoid convoluted phrases which can lead to misinterpretation. The author recently ran across the sentence: "This method will detect not less than 0.02 ppm of. . . ." in a method. The use of "not less than" was momentarily confusing until it was realized that it was equivalent to "more than," and the statement was meant to define a detection limit.

10. Calculations - Give the equation(s) necessary to calculate the results of the analysis, including the units of all variables and the units of derived results. If the equations are not straightforward, indicate how they were derived.
11. Statistics - Give all information available on the precision and accuracy of the method in summarized form. Refer, if necessary, to the source of the method and to any validation studies that were run in the laboratory.
12. Quality Assurance - This section should indicate what reference samples are available and how frequently they should be analyzed.
13. Comments - Any special comments or remarks that may add to the understanding of the method or interpretation of results should be added here.

Modification of Analytical Methods

In the course of normal operations of a laboratory running routine analyses, it will often be found that minor modifications of a method are desirable for various reasons. This could take the form of a change in sample size, slightly different pH, or similar slight change in one or more variables. It is tempting to simply permit these modifications without validation or subsequent documentation.

All such modifications should be subject to some type of validation. This could take the form of simply running several samples by the modified and unmodified methods with comparison of results, if the modification is truly minor and judged by the supervisor to be unlikely to affect results. On the other hand, major modifications which might be suspected of affecting results should have a full validation study, as described earlier in this chapter.

Regardless of how minor the modification, if it is adopted as part of the method, it must be documented and become a part of the written version. This can be done by addendum to the written method, with notes in the procedure referring the reader to the written modification. The written modification document should indicate clearly what the modification consists of, the validation data or reference to a validation study which may be documented elsewhere, and the date at which the modification was adopted as part of the authorized method.

AUTHORIZATION OF ANALYTICAL METHODS

As discussed earlier, the set of analytical methods used in the laboratory is in effect the communication medium by which management controls the way in which the work is done. As such, management must exert some form of control over the methods which are used. In many laboratories, this is accomplished in an informal way, if at all. Many managers have a simple faith in the competence of their laboratory supervisors and analysts, and in the overwhelming majority of cases this faith is well placed. However, in the modern world, where serious consequences may result from poor quality data, much more emphasis is being placed on the duty and responsibility of the manager to ensure accurate analytical data, and the manager may find himself held personally liable for the consequences of poor data.

Therefore, it is incumbent on laboratory managers to become familiar with the methods that are actually in use in their laboratories, with all of the limitations and variability that may be expected. For quality assurance purposes, it is necessary that documentation exists which demonstrates that the method has been authorized for use by the manager responsible for the quality of the data. This may be accomplished by establishing a procedure whereby the manager reviews all pertinent references and in-house documentation and completes an "authorization" form, which permits the method to be used in the laboratory. A copy of the form should be attached to the written analytical method. The original should be kept in the Quality Assurance Director's files. An example of such a form is given in Figure 4–2.

The use of such a form (and the procedure of method review by the manager) together with a directive which states that only *authorized* analytical methods may be used, will prevent the encroachment and use of untested, unproven methods in analytical operations. It should be pointed out to analysts that the use of unauthorized methods will not only jeopardize their positions, but may make them legally responsible for the results.

METHOD AUTHORIZATION FORM

Title of Method: Date:

Analyte (substance being analyzed for): Submitted by:

Matrix (material(s) being analyzed):

Brief description of scientific basis of method:

Brief description of validation study:

Results of validation study:

a. Accuracy (standard error of residuals, or evidence of bias):
b. Precision (standard deviation or coefficient of variance):
c. Interferences:
d. Applicable concentration ranges:
e. Limit of detection (give basis):
f. Results of ruggedness test:

Validation study performed by:

Validation study data in Notebook No.:

Approved ______________________________

Name Title

Date ______________

11/8/83

Figure 4-2 Method Authorization Form

REFERENCES

1. Mandel, J. and Linnig, F. J., Study of Accuracy in Chemical Analysis Using Linear Calibration Curves, *Analytical Chemistry,* 29: 743-749 (1957).
2. Youden, W. J. and Steiner, E. H., *Statistical Manual of the Association of Official Analytical Chemists,* Washington, D.C.: Association of Official Analytical Chemists, 1975.

5
INSTRUMENT AND EQUIPMENT CALIBRATION AND MAINTENANCE

CALIBRATION

Almost all discussions of quality assurance agree on one subject: the need for frequent calibration of instruments and other equipment. The reason for this is quite obvious. Calibration is performed to assure that the equipment is operating correctly and thus that one possible source of error is under control. If one looks closely at the operational aspects of equipment calibration, however, the concept begins to get a bit fuzzy around the edges.

Instrument Calibration

Calibration implies comparison with a standard, i.e., a material with a known value or property, and hence an expected instrument or equipment response. Problems arise from the choice of standard and from the type of instrument response expected or evaluated. Standards can range from Standard Reference Materials, certified by the National Bureau of Standards, and thermometers certified by NBS, through equipment certified by manufacturers (glassware, class *S* weights, etc.) to reagent chemicals which are certified as to purity by the manufacturer. The choice of standard will depend on the required accuracy of the analytical methods using the instrument or other equipment, and to some extent on cost. NBS Standard Reference Materials, for example, are very expensive and hence used infrequently, or only when necessary.

Another difficulty arises from the fact that the term ''instrument'' covers devices which vary from the relatively simple, e.g. a refractometer, to the very complex, e.g., a gas chromatograph/mass spectrometer. Distilled or de-ionized water will serve to calibrate a refractometer, a film of polystyrene will serve for an infrared spectrophotometer and a solution of potassium permanganate may be used for a UV/visible spectrophotometer. However, a device such as a gas chromatograph, or atomic absorption spectrophotometer, is in reality a *system,* composed of a number of separately functioning components, and does not belong in the same class as the simpler "instruments." For example, a gas chromatograph may consist of an automatic sample injector, an injection port with associated temperature control, a carrier gas flow system, a chromatographic column with associated temperature controls (constant or programmed), a detector with its temperature control and electronic circuitry, a strip-chart recorder and/or an electronic integrator. Theoretically, each of these components could be separately calibrated, but the practical difficulties of doing this are obvious.

Instead, calibration is made a part of the analytical method, by preparing a calibration curve using solutions of known concentration and measuring instrument response to these "standard" solutions. In the not-too-distant past, such calibration curves were manually generated, and the concentration of unknown solutions derived from samples were calculated by visual interpolation from the graph. This is still done today in laboratories using older instruments, but the trend in modern instrumentation is for the whole process to be carried out electronically under the control of the ubiquitous microprocessor. One injects or aspirates into the instrument aliquots of the standard solution and the microprocessor essentially calculates the least squares straight line for instrument response versus concentration. Using the calculated equation, the microprocessor then calculates the concentration from instrument response on subsequent injection or aspiration of unknown solutions. The mathematics of the process is described in the following section.

Linear Calibration Curves

We will first consider *linear* calibration curves, i.e., curves which follow the straight line equation:

$$y = mx + b \tag{5-1}$$

where y represents instrument response, and x is the concentration of the standard solution. This is strictly analogous to equation (4–4) of Chapter 4 and the same method is used to calculate m and b. The parameters:

$$S = \Sigma x \qquad Q = \Sigma x^2$$
$$Y = \Sigma y \qquad L = \Sigma y^2$$
$$P = \Sigma xy \qquad D = NQ - S^2$$

are calculated (where N is the number of standard solutions). The slope m and intercept b are then given by:

$$m = (NP - SY)/D \tag{5–2}$$

$$b = (QY - SP)/D \tag{5–3}$$

Incidentally, a "blank" or zero concentration solution made up of all reagents used may be included as one of the solutions.

The instrument response will have a random error component such that the measured responses will not all fall on the least squares line. If we call $\hat{y}_i$ the response corresponding to the concentration x_i given by $mx_i + b$, and y_i the actual measured response, then:

$$y_i = \hat{y}_i + e_i = mx_i + b + e_i \tag{5–4}$$

where e_i is the random error associated with the instrument response at the time the measurement was made. It can be assumed that these deviations from the fitted line are normally distributed and hence can be described by a standard deviation s_e which is called the "standard error of estimate." The easiest way to calculate s_e is to calculate e_i for each point, using equation (5–4) and then calculate s_e from:

$$s_e = [\Sigma e_i^2/(N - 1)]^{1/2} \tag{5–5}$$

If an unknown solution is now measured, giving an instrument response y_1, then its concentration will be given by:

$$x_1 = (y_1 - b)/m \tag{5–6}$$

and its standard deviation by

$$s_x = s_e / m \tag{5-7}$$

Equation (5–7) is not rigorous, because it does not take into consideration the variation in m and b, but it is sufficient for most practical applications.

What are the implications for quality control of the above analysis? One of the major objectives of quality control as applied to instrumentation is to detect instrument malfunction before it becomes serious enough to degrade the quality of results. It is widely believed that if standards are run, and calibration curves prepared each time the instrument is used, that this is sufficient. However, it is quite possible for an instrument or a component of an instrument system to slowly degrade with time, yielding progressively less accurate data. Of course, this will be detected if check standards or spiked standards are used and $\overline{X}$-charts kept. If possible, it is recommended that QC charts be kept of the slope, intercept, and standard error of estimate as well. Unfortunately, manufacturers of instruments in many cases have not programmed the microprocessor to automatically print out this data when calibration curves are run.

Equipment Calibration

Equipment other than instrumentation also requires calibration for good quality assurance. By equipment is meant all devices, other than instruments, used in analytical procedures. Instruments may be defined as those devices which respond to a property of matter. Equipment includes glassware, ovens, furnaces, automatic pipettors, and a myriad of special devices, e.g., viscometers, flash point apparatus, thermostatted water baths, etc. As in the case of instruments, calibration implies comparison against a standard. Oven temperatures are measured with thermometers which can be calibrated against NBS certified thermometers. Viscometers and flash point apparatus can be calibrated against materials of known properties, etc. Ordinary volumetric glassware is not usually calibrated since calibration is done by the manufacturer and is part of the guaranteed specification against which it is purchased. However, twice in the author's career of more than forty years he has discovered burets which were mislabeled, i.e., had one number missing. In one case, a 50 ml buret had a number missing in the 40–50 ml range, being labeled 43, 44, 46, . . . , 50. Titrations which required more than 45 ml were misread, while those at lesser volumes were read correctly. A defect such as this can be very difficult to spot in a casual glance at the calibration marks and the defective item can be used for a long time before the situation is discovered.

Frequency of Calibration

The frequency with which instruments and equipment should be calibrated is a function of many variables, e.g., the nature of the instrument and its ruggedness, frequency of use of the instrument, its environment, etc. Calibration schedules can be set up according to frequency to use or according to elapsed time. If an instrument is not used frequently, or is of a very sensitive nature, it may require calibration each time it is used. With instruments e.g., gas chromatographs, high-performance liquid chromatographs, atomic absorption spectrometers, and the like, calibration via standard curve may be an integral part of the analytical method. A visible spectrophotometer, however, may only require calibration once a month, while pH meters should be calibrated each time they are used. In general, scientific judgment should be used, with the general principle that the more often the calibration the better.

Calibration frequency may depend on the importance of the sample. If the analytical result may become part of the evidence in a legal dispute, or if costly decisions will be made based on its accuracy, all instruments and equipment used in the analysis should be calibrated before and after the analysis.

Calibration of Common Laboratory Instruments and Equipment

The following is a brief summary of calibration methods and guidelines for frequency of calibration for the more common instruments and equipment used in the analytical chemistry laboratory.

Analytical Balances. Should be checked with high quality (at least Class S) 50 mg or 100 mg weight. Calibration frequency is dependent on frequency of use, daily or weekly if frequently used.

Volumetric Glassware. If the method requires the highest accuracy, Class A glassware should be used. Class A glassware is guaranteed by the manufacturer to meet NBS specifications and does not need calibration. Other types of glassware may be used, if lower accuracy can be tolerated, but should be stored separately from Class A. Glassware may be calibrated gravimetrically using water, or in the case of small volumes, mercury.

Ovens. Thermometers used in ovens should be NBS certified or calibrated against NBS certified thermometers. Oven temperature should be checked daily.

Furnaces. Calibration of muffle furnaces is difficult. Fortunately, for most purposes, accurate calibration is not necessary. Optical pyrometers, or high temperature probes, may be used, but usually require that the oven door be at least partially open. High melting inorganic salts may be used, choosing two which bracket the nominal temperature of the oven. Calibration on a semi-annual basis is usually sufficient.

UV/Visible Spectrophotometer. Various standards are used. NBS certified glasses are available. A solution of 0.0400 gm K_2CrO_4/liter in 0.05M KOH has absorbance peaks in both UV and visible regions. A $KMnO_4$ solution may be used to check resolution of the peaks at 526 and 546 nm in the visible region. Test kits are also available in sealed test tubes for calibrating simple visible spectrophotometers for both wavelength and absorbance. Calibration frequency will depend on frequency of use. If used daily, calibration should be done at least biweekly.

pH Meters. Calibration is carried out with standard buffer solutions. pH 4 and 7 buffers are frequently used, with pH 10 included if frequent alkaline pH's are encountered. If the meter is frequently used to measure pH of solutions that are particularly "dirty," e.g., wastewater which may contain high concentrations of solid matter or oil and grease, it may be found that two or even three buffer solutions are not sufficient to detect a non-linear response of the glass electrode, which may have become fouled. In this case, a series of buffer solutions, covering the pH range from 2 to 11 may be prepared and a response curve determined, plotting pH as read on the meter against nominal pH of the solutions.

Table 5-1 indicates how such a series of solutions may be prepared, using various quantities of two reagent solutions.

pH meters should be calibrated each time used, or on a daily or twice daily basis if frequently used. The full-range calibration described above may be done monthly or biweekly, if the meter is frequently used in dirty solutions.

Infrared Spectrophotometers. A thin film of polystyrene is the most common standard used for IR spectrophotometers. It is stable and contains a rich mixture of absorption peaks. Transmission and wave number of the peaks can be compared with previous runs to detect any deterioration in performance.

Atomic Absorption Spectophotometers. These instruments fall in the class of those which are generally calibrated each time used, as described previously under "Linear Calibration Curves." Certified standard solutions (1 gm/L) for various elements are available from various chemical supply houses. These solutions have been checked against NBS standards.

Conductivity Meters. These instruments may be calibrated against a standard KC1 or NaC1 solution. Frequency of calibration will vary depending on frequency of use, but should be done at least weekly, if heavily used.

Table 5-1

pH	SOLUTION *A* (mL)	SOLUTION *B* (mL)
3.0	15.89	4.11
4.0	12.29	7.71
5.0	9.70	10.30
6.0	7.37	12.63
7.0	3.53	16.47
8.0	0.55	19.45

Solution *A* - 0.1M citric acid
Solution *B* - 0.2M disodium phosphate

Dissolved Oxygen Meters. This is a good example of an instrument for which there is no method of preparing an acceptable standard solution. Calibration is performed by comparison with another method of acceptable accuracy. In this case, a suitable sample of water saturated with oxygen is analyzed by both the Dissolved Oxygen Meter and the Winkler iodometric titration metod. The result of the Winkler method is assumed to be the "true" value of the oxygen concentration. Calibration should be done at least monthly if the instrument is used frequently.

Gas Chromatographs and High Performance Liquid Chromatographs. These are instruments which fall in the class of those which are calibrated via calibration curve with each batch of samples run. Calibration standard solutions should be prepared with materials of high purity, (>99.9%), which are commercially available in most cases. Calibration curves developed for frequently run analyses should be periodically checked against previous runs to detect any changes in instrument response.

MAINTENANCE OF INSTRUMENTS AND EQUIPMENT

Instrumentation is the most potent amplifier of productivity available to the manager of the analytical laboratory. Even though the cost of instrumentation has increased substantially over the years, it has not equalled the rate of increase of labor cost which has always been the major cost of analytical work. This has made the investment in more and more sophisticated equipment an attractive method of allocation of financial resources. What is not often as well appreciated by management, however, is the return on investment of proper maintenance of expensive equipment.

In the not-too-distant past, instrument and equipment maintenance was pretty much left up to the analytical laboratory personnel. Most chemists had sufficient knowledge of electricity, vacuum-tube electronics, and optics to repair equipment with the aid of a vacuum-tube voltmeter, soldering iron, and a few hand tools. Furthermore, the principles of operation of the instrument were sufficiently simple to be easily understood. However, modern equipment has become much more complex, particularly in its reliance on solid-state, digital electronics, and few chemists today are sufficiently knowledgeable to perform any but the simplest repairs on such instruments. The new electronic and microprocessor systems have the major advantage of being much more reliable and less likely to require maintenance than the old vacuum-tube technology, but have the disadvantage of the possibility of very slow degradation in performance which is not detectable except by frequent calibrations.

One result of this situation has been the creation of a new profession, namely "instrument technician." Very few laboratories have sufficient instruments to warrant hiring a person full time for instrument repair and maintenance, so these services are usually supplied by either the manufacturer or a firm which specializes in instrument repair. As with all labor intensive services today, the cost of having an instrument repaired has skyrocketed. Typical hourly rates are $40–$70, with travel time and expenses often included, so that unless a local firm can be found to do the work, a major expense is easily incurred. In addition to out-of-pocket cost, the loss of productivity due to instrument downtime can be considerable. "Service contracts" are available from most instrument manufacturers and are probably a good idea, especially during the first year of a new instrument's life.

Instrument manufacturers are aware of this problem and are responding in several ways, e.g., using components and circuitry with proven records of reliability. The analytical chemistry community is a close-knit one, and a manufacturer whose instruments and service departments have a poor reputation has a major marketing problem. Many manufacturers are also adopting a modular approach or using plug-in printed circuit boards which permit fast diagnosis and repair. Telephone consultation with the manufacturer's service department can also be useful in effecting minor repairs and diagnosis. The areas which still need improvement, however, are design of instrumentation to facilitate removal of components, built-in diagnostic tests for locating faulty circuits, and better documentation from the point of view of explanation of circuitry and fixing equipment.

Proper preventive maintenance of analytical equipment begins with the environment in which it is placed and used. Modern instruments are invariably sensitive devices in that they are designed to give maximum response for minimum input of detected material. For this reason they are vulnerable to environmental influences, e.g., dust, vibration, excessive heat, humidity, corrosive fumes, radiation, etc. If at all possible, instruments should be kept and used in a separate room or rooms set aside solely for this purpose. At a minimum, such an instrument room must be air-conditioned in the summer and heated in the winter. It is desirable for the instrument room to have a heating, ventilation and air-conditioning (HVAC) system which is independent of the rest of the building, and be kept at a positive air pressure with respect to the environment outside the room. This will guard against the accidental intrusion of dust, fumes, and vapors from other operations in the building or outside the building. With the very low detection limits achievable with some of today's instrumentation (parts per billion or less), it is almost mandatory to isolate these instruments against outside contamination for reproducible results.

Personnel should be instructed in the proper use and care of instruments. Persons who are not authorized to use instrumentation, i.e., who have not been specifically instructed and trained, should be prohibited from using or handling costly instruments. Most people will approach a new instrument with caution, being mindful of the cost of the equipment and the dire consequences of causing a failure, but there are those few who will rush in where angels fear to tread. Indiscriminate knob twiddling, switch snapping, and button pushing should be vigorously discouraged.

Items which need frequent replacement in an instrument should be purchased before they are needed and kept in stock. There is nothing more frustrating than instrument downtime caused by waiting for an inexpensive part, e.g., a light source, to arrive by mail or courier. A little forethought and planning when a new instrument is purchased can be very effective in avoiding costly shutdowns.

All modern instrumentation relies on electricity. Care should be taken that the capacity of the line to an instrument is not exceeded. Even if the line is not loaded sufficiently to blow a fuse or circuit breaker, heavy loading will cause excessive warmup times, low voltages, or voltage surges when instruments on the same line are turned on or off. One sign of excessive loading may be a microprocessor which loses memory, or which cannot be programmed properly, or is not following a presumed program. Sometimes this type of malfunction can be cured by turning the instrument off, disconnecting some of the load on the line, waiting a few minutes to allow the electrolytic capacitors in the instrument to drain, and then turning the instrument on again. If this cures the problem, it is a good sign that the line is excessively loaded.

All modern instruments are equipped with three-prong plugs, the third prong serving the purpose of grounding the chassis and outer case of the instrument. This is done for two reasons: safety, so that the outer case or chassis cannot become "hot," and to shield the inner electronics from external electrical fields, e.g., those produced by sparking electric motors or automotive ignition. Although all instruments today have three-pronged plugs, many laboratories do not have three-hole receptacles. Use is often made of an "adapter" for converting a two-hole recepacle to accept a three-prong plug, the third hole being connected to the face plate screw which connects to the box surrounding the receptacle. The assumption is made that the box is grounded, but this is often not the case. In fact, the author has often found three-holed receptacles in which the third hole was not grounded, especially in older laboratories. Fortunately there are inexpensive devices available for checking receptacles to see if they are properly grounded. Any time an instrument is found to behave erratically, especially when other instruments or electrical equipment are turned on or off, inadequate grounding should be suspected.

Another source of instrument malfunction can be voltage surges caused by nearby lightning strikes which are not close enough to burn out lines or equipment but cause ground currents of large magnitude. These can in turn cause voltage "spikes" in the kilovolt range to appear on the lines. The lifetime of these voltage spikes may be only microseconds, but solid-state devices, e.g., transistors, integrated circuits, and diodes can be destroyed easily in this timeframe. Surge protection may be built into the circuitry of the instruments, or protective devices may be purchased for relatively little cost. The latter are based on the use of metal oxide varistors (MOV's) which exhibit a resistance inversely proportional to the applied voltage and have a reaction time of microseconds, unlike ordinary fuses or circuit breakers which do not react rapidly enough to protect solid-state electronic circuitry. MOV's can also be installed in the line to which the equipment is plugged, or in the service entrance line, by a qualified electrician.

Wherever electricity is widely used, safety should be of prime importance. Modern equipment, properly grounded, with fuses and circuit breakers, will offer excellent protection against the hazard of electrical shock as long as the instrument case is not opened. However, power cords should be inspected periodically for cracking, fraying, etc., and replaced at the first sign of deterioration because of the possibility of ignition and fire. Fire extinguishers used in the instrumentation lab should be of the halocarbon type to avoid damaging the instruments should they have to be used.

6
DOCUMENTATION FOR QUALITY ASSURANCE

Quality control as defined earlier in this book consists of those operations, technical in nature, by which the quality (accuracy) of the data is measured and controlled within known limits. This quantification of accuracy limits is achieved by controlling the various elements of the system by which the data are produced: instruments, reagents, and operations of the analysts. In a certain sense, it is simply good scientific practice, codified and reduced to a set of rules for carrying out chemical analyses.

If analytical chemistry were carried out in a vacuum, as a purely intellectual pursuit, QC would be sufficient. However, analytical chemistry is always performed for a purpose, i.e., it is an information-generating process and its output is of real economic value to the outside world. For this reason, QC by itself is not sufficient since the outside world of clients, purchasers, accreditors, lawyers, judges, etc. needs to be assured of the quality of analytical data. Quality assurance arose as an answer to this problem and has now been extended to assurance of the security and traceability of the data as well as quality control. Quality assurance is an outwardly directed activity by which the laboratory convinces outside auditors of the quality of the information it produces.

By its very nature, quality assurance implies documentation. It is not sufficient to carry out quality control operations; it is necessary to document the fact that they were carried out, and by whom, and when, and with what results. For example, it is good scientific practice to calibrate a pH meter each time or day it is used by means of buffer solutions, and this is also good quality control. Quality assurance requires that the fact that the meter was calibrated shall be recorded in a notebook, along with the identity of the person who did the calibration, the date, and the results.

The remainder of this chapter will explore the kinds of documentation necessary for a good quality assurance program.

NOTEBOOKS

For some reason, scientists are addicted to taking down data on the backs of old envelopes, odd pieces of paper, or any other writing surface which happens to be handy. It almost appears that they are more interested in the process by which the data are generated than they are in the data themselves. Needless to say, this philosophy is exactly counter to good scientific or commercial practice, and especially to quality assurance. Quality assurance demands that primary data (also called raw data) be recorded in as permanent a fashion as is reasonable, in such a way as to facilitate traceability of derived results, and with reasonable precautions against future falsification of data.

Type of Notebook

All notebooks used to record primary data should be bound and have consecutively numbered pages. If notebooks which are not bound are used, e.g., looseleaf, or spiral bound, pages may easily be lost, and the possibility of fraudulent tampering with the data has not been safeguarded against. It is common practice in many labs today, especially where multiple routine analyses are run, to use printed or mimeographed "work sheets" or "data sheets" for analyses in which a lot of data are recorded per analytical result. This procedure is not good quality assurance, but is an efficient system of data collection. One possible compromise is to have the work sheets bound into books, and to number the pages consecutively as they are used.

Control of Notebook Distribution

To avoid loss or misplacement of notebooks, a system should be established for the issuance of notebooks and the storage of completed notebooks. All notebooks should be consecutively numbered and a master file set up containing the number of the book, type (if more than one type or size is used), date of issue, name of person to whom issued, date of return of notebook, and storage location of returned notebook. Responsibility for notebook control should be in the hands of one person. This may be the Quality Assurance Director, but may also be a secretary or one of the analysts. Notebooks do not necessarily have to be returned as soon as they are completed—some analysts prefer to retain notebooks for a while for reference—but the analysts should know that they are personally responsible for notebooks issued to them until they are returned for storage.

In general, it is good practice from a quality assurance standpoint for each analyst to have his or her own notebook. While this is perfectly feasible in many laboratories, in others it is less than desirable from the point of view of efficiency of operations. In some laboratories where large numbers of samples are run through an analysis simultaneously, it may be more efficient to keep different notebooks for different operations. For example, one book might log sample weights, a second titration volumes, etc. The book would then be logged by whoever happens to be performing that part of the operations on a given day. From a traceability standpoint, it may also be more efficient to keep data according to sample type, e.g., one book for food analyses, another for water, etc. If systems such as these are in use in place of individual notebooks, a document describing the system in use should be in the Quality Assurance Director's file.

Data Entry

Data entries in the notebook should be subject to rigid regulations. All data entries must be made in ink of the non-erasable type. Errors must not be erased, eradicated or "whited out." Erroneous entries are cancelled with a single line drawn through them, in such a manner as not to obscure the nature of the error. A brief note in the margin near the cancelled entry should explain the reason for the cancellation.

Blank pages or substantial portions of pages with no entries should be marked with a large X to indicate that they were deliberately or unintentionally left blank and are not available for addition of new data at a later date.

All entries in a notebook should be dated and signed by the person making the entry. If the entry covers more than one page, each page should be dated and signed. The most desirable signature is a full, legal signature but for some reason analysts seem to prefer using initials. Initials *per se* are not acceptable, because it is quite possible for two persons in an organization to have the same set of initials. Since the major reason for this requirement is to trace results back to the individual who ran the analysis, some system of unambiguously identifying the person who made the entry must be used. In some organizations, unambiguous employee identification numbers are issued for payroll and other purposes. These may be used, if accompanied by the person's initials as a sign that the entries were not forged. Another scheme is to list on the front cover of the book all employees who have access to the book, identifying them by both name and initials (provided no two persons with the same initials use the same book).

In addition to signatures of those making data entries, the laboratory supervisor or his assistant should sign the book periodically. This should be done at intervals, e.g., every twenty-five pages, and should be accompanied by a statement such as "Read and understood, pp __ through __ John Q. Doe." A rubber stamp with the phrase may be used, but the signature must be genuine. The purpose of this is to signify the supervisor's responsibility for seeing that the book is being properly used, and establish that he or she is periodically reviewing that such is the case.

ANALYTICAL METHODS

In Chapter 4, analytical methods and the format in which they should be written were discussed. All analytical methods should be in written form and kept where they are available to all personnel who have occasion to use them. This is most conveniently done by means of an "Analytical Methods Book" or manual, which should be kept in the laboratory itself where it can be conveniently referred to by the workers. Such a book is best kept as a looseleaf binder, since it should contain only those methods currently in use, to avoid confusion regarding which method should be used and to enable easy removal of methods no longer used. On the other hand, at some place in the organization, preferably in the Quality Assurance Director's office, a historical file containing *all* methods, current and obsolete, must be maintained for traceability purposes.

Chapter 4 discusses the recommended procedure by which an analytical method becomes authorized for use in the laboratory. Authorization of a method by management should always be based on validation of the method, and the procedure for validating a method is necessarily based on its source. Methods which are decreed "official" by government agencies such as the EPA, which refuse to accept data based on other methods, need not necessarily be subjected to a validation study, since responsibility for the method then rests with the agency involved. On the other hand, methods which have been adopted as "standards" by voluntary standard-setting organizations, e.g., ASTM, ANSI, or AOAC, may need only a few samples run to be sure that the method as written is thoroughly understood by the analysts and capable of turning out acceptable data. Methods adopted from the literature and methods devised within the laboratory should be subjected to a much more elaborate validation study, as discussed in Chapter 3.

All data pertaining to a validation study should be entered into notebooks maintained in the same manner as described in the previous section of this chapter. These notebooks should be kept separately from those used for entering raw data collected in the course of routine analyses using already authorized methods. The results of a validation study should always be signed by at least one other competent, degreed analyst in addition to the person who performed the study, using the phrase "Read and understood by ______."

All authorized analytical methods should be given an official laboratory method number. Any system of numbering may be used, from a simple sequential scheme to an alphanumeric designation which may distinguish methods used in one lab, or on one substrate, from another. Regardless of the system used, one cardinal principle should apply: No more than one method to a number and no more than one number to a method. The reason, of course, is that the method numbers are used for traceability purposes, and if the principle is violated there will be confusion over which method was actually used at a given time. It should be borne in mind that methods which differ from each other in only minor ways, e.g., the use of different sample sizes for different substrates or different analyte concentrations, need not be given separate numbers, since these variations can be described in the body of the method. Also, a new number does not need to be issued when a method undergoes a minor modification. It is only necessary to indicate the date of the revision on the modified method. The original method will be available in the historical file.

The distinction between a "new" method and a "revised" method can sometimes be a source of confusion. In general, any method which is based on a different measuring principle, i.e., has a different scientific basis for measurement, must be regarded as a new method and given a new number. Methods which involve minor changes in techniques, e.g., different sample weights, or dilution factors, may be considered revisions of existing methods. Such revisions should always be validated before being adopted as authorized, since it often happens that a so-called minor revision can lead to unexpected errors in analysis.

REAGENTS AND REAGENT SOLUTIONS

The purity of reagents and the composition of solutions prepared from reagents are, of course, of paramount importance in analytical operations. One of the major sources of "outliers," i.e., results which are outside the limits of random error causes, is contaminated reagents or reagents which have been incorrectly prepared. Preventing these errors, or at least providing a means of identifying them in suspect results thus becomes one of the objectives of a good quality assurance program.

Purchasing reagents with the necessary degree of purity is well understood by managers of analytical laboratories, and manufacturers of reagents have done an admirable job of providing reagents of varying grades to suit the needs of the analyst. The American Chemical Society, the U.S. Pharmacopeia, and other organizations have assisted by setting standards which permit manufacturers to label reagents: ACS Grade, U.S.P., Analytical Reagent and the like, with published maximum contamination limits. Although mislabeling of manufactured reagents has occurred, it is so rare that it is seldom even considered as a possible cause of erroneous results.

The problems with reagents arise after they have been received in the laboratory. As soon as the container is opened, the possibility of contamination is present. Bottles which are not tightly resealed expose the contents to air with the possible loss or pickup of moisture, absorption of carbon dioxide, or absorption of contaminating vapors which may be present. Material may be removed with a contaminated spatula, especially by untrained technicians who often do not understand the extreme care needed to prevent contamination.

Purchased reagents should always be labeled with the date received and the date the bottle was opened. In addition, an expiration date can be added, if the material is likely to degrade over a period of time. While such labeling will not prevent the contamination problems described above, it will alert the experienced analyst to possible reagent problems at the time the material is used or when a suspicious result has been obtained. Separate labels should be used for this purpose, i.e., the relevant dates should not be inscribed on the existing manufacturer's label, because such inscriptions are often difficult to read, or may obscure important information on the label. Labels are commercially available printed with the appropriate legend, with blank spaces for insertion of dates.

Reagent solutions prepared in the laboratory should also be labeled with the date of preparation, the concentrations of active ingredients, and an expiration date. Even solutions which are known to be stable should be provided with an expiration date, beyond which they should not be used. The reason is that solutions are generally used frequently, opened and closed again, with more or less airtight seals. The possibility of contamination, evaporation of solvent, leaching of impurities from containers, increases with time. No reagent solution more than six months old should be used in an analytical laboratory.

One practice which is prevalent in many laboratories, and which should be vigorously discouraged, is pouring unused portions of solutions back into the reagent bottle. An analyst will often pour the reagent out into a beaker, extract an aliquot via pipet or by filling a buret, and then pour the remainder of the solution back into the bottle. The rationale for this behavior is conservation of costly reagents and time of preparation, but the possibilities for contamination are obvious.

Every laboratory should have a Reagent Preparation Notebook. This should be a bound notebook, kept in the manner previously described. When a reagent solution is prepared, the analyst should record in the notebook the date, the actual quantities of weight and volume used, and his or her signature. This will be of inestimable assistance in tracing possible sources of error in analyses. Analysts are human beings and are quite capable of preparing a 10% solution, when a 1% solution was desired. If forced to record actual quantities used, such errors are more easily detected.

SAMPLING AND SAMPLE HANDLING

Sampling Methods Manual

It is a truism in analytical chemistry that an analysis is no better than the sample upon which it was performed. In other words, the most accurate method possible will give useless results if the sample is not representative of the lot of material for which the composition is required. Chemical analyses are always performed for a purpose, which is to generate information regarding the chemical properties of a more or less well-defined mass of material. Usually this mass of material is larger than that required for the analysis (or in many cases the entire mass is not available for analysis) and so care must be taken that the sample used for analysis be as representative as possible of the whole, at least as far as chemical composition or properties is concerned.

Methods of achieving representative samples are varied, depending on the nature and amount of material available, and are beyond the scope of this book. However, quality assurance dictates that the same documentation be undertaken for sampling as for analysis, especially in those cases where obtaining the sample is the responsibility of laboratory personnel.

Of prime importance is the use of written procedures for sampling, collected in a Sampling Methods Manual. The manual should be kept in the laboratory where it will be available to all personnel involved in obtaining samples. Only current sampling methods should be in the manual, with a historical file of current and obsolete methods kept in the Quality Assurance Director's file.

A distinction should be made here between the concept of "sampling" or the obtaining of a small portion of a larger mass which is representative of the mass, and "sample reduction" which is ensuring that the sample used in analysis is representative of the sample presented to the analyst. "Sample reduction" is generally written up as part of the analytical method itself, while "sampling" is more often a separate operation, often carried out by personnel other than the analyst. It is the laboratory supervisor's responsibility to see that personnel used for sampling be thoroughly instructed in the techniques used, and hence the need for a sampling methods manual. If the sampling takes place at a remote location, a copy of the manual may be kept in the vehicle used to transport personnel and equipment to the place where the sample is to be taken.

Sample Labeling

As soon as a sample is taken, a label should be prepared and attached to the sample container. Various systems for accomplishing this are used, depending somewhat on the circumstances under which the sample is taken. For example, sample containers may be numbered, and a log sheet filled out which contains pertinent information logged against the sample number. In other cases the container has the label attached prior to sampling and the information is entered at the time the sample is taken. The important principle here is that the information should be entered when the sample is taken, and not before or after the sample is taken.

The minimum information required on a sample label should be:

- Sample designation – a number or alphanumeric symbol unique to the sample.
- Description – a brief description of type of sample, e.g., "wastewater," "reactor bottoms," "soil," etc.
- Location – an unambiguous designation of the place the sample was taken
- Time and date when taken
- Sampler – full name of person who took sample

Ancillary information, such as temperature or the results of spot tests performed at time of sampling (pH, chlorine residual, etc.) may also be entered on the label if these may be important to the analysis or interpretation of results.

One important point here which is obvious but still frequently overlooked is that the label should never be attached to the cap or lid of the sample container, but only to the container itself. Lids and containers can be easily interchanged, leading to mislabeling of the samples. Labels should be filled out in ink, using an ink which is water resistant.

A log sheet describing the samples taken should be filled out and signed by the sampler, even if all of the above information is entered on the labels. On return to the laboratory, the samples should be delivered directly to a responsible person who in turn signs the log sheet which is then filed. Only in this way is documentation maintained that the samples have, in fact, been taken and received by the laboratory personnel.

Sample Log-In

Immediately on receipt in the laboratory, samples should be logged into the laboratory system. At this point, the sample should be given an unambiguous laboratory number, even if it already has a sample number bestowed on it by the sampler or client. The sample log-in procedure is a very important step in the "chain of custody" sequence and one in which many laboratories are vulnerable to legal challenge. All samples should be logged-in to the system on the day they are received. The sample log should contain the following information:

- The laboratory-designated sample number. Various systems are used, but the important principle is that the number be unambiguous, i.e., unique to the particular sample. One system is to use the last digit of the year, followed by three digits designating the day of the year (which can be obtained from many calendars), followed by a consecutive number. Thus 30247 would designate the seventh sample received on the 24th day of 1983. The advantage of this system is that the date of receipt is immediately obvious in the sample number. This numbering system will recycle every decade, but this should cause no problems in most laboratories.
- Name of the client and/or his company or organizational unit.
- Client's sample designation, i.e., the sample number or description used by the client for his records.
- Sampler's name, if known. If not, the fact that the sample was taken and submitted by the client should be indicated.
- Nature of the matrix—water, soil, organic liquid, waste solid, etc.
- Analyses requested and method to be used, if this is specified by the client or other circumstances.

- Sample storage location. Most samples are not analyzed immediately upon receipt at the laboratory. They must be stored somewhere until they are able to be fitted into the laboratory operation. Sample storage areas should be set aside and so designated to guard against sample loss and/or mixup. For small samples, a series of shelves may be provided, divided up into separate cubby holes, and each cubby hole given a numerical designation. This designation can be entered on the sample log-in sheet and will provide the analyst with information on sample location when he/she is ready to analyze the sample. Larger samples may have to be stored on the floor, but they are usually easier to find.
- Special storage conditions. Samples may require refrigeration, freezing, storage under inert atmosphere, in the dark, etc. This information should be entered on the log-in sheet. It may also be of advantage to log an expiration date after which the sample is to be discarded. (This may take the form of a set number of days after analysis.) This will tend to prevent the accumulation of large numbers of samples which are no longer of interest.

Sample log-in in the small laboratory is conveniently carried out in a ledger-type notebook, and the usual precautions against future tampering with entries should be observed. The book should be bound, with all entries made in ink, dated, cancellations made with a line drawn through the entry, and entries signed by the person making the entry.

In large laboratories handling many samples each day, the modern trend is toward computer log-in. Here the computer assigns the sample number, requests all pertinent information, and may actually print the labels to be attached to the sample. The database thus generated can be used as the basis for a sample management system which will track the sample though the laboratory, do all calculations, and generate the final report. (See Chapter 7.) If computer log-in is used in the laboratory, it is important to remember that computers are not immune from disasters such as "head crashes" where information can be lost. This necessitates frequent data back-up on auxiliary disks and a hard copy printout at least once each working day. This printout will also serve as an excellent management tool for estimating work load and work in progress.

INSTRUMENT CALIBRATION AND MAINTENANCE

For traceability purposes, all calibration and maintenance work done on instruments should be documented. If an analytical result is questioned, the fact that a major component of an instrument showed signs of malfunction or needed to be replaced at about the time the sample was tested, could be an important clue to the cause of a poor result. On the other hand, documentation that an instrument was calibrated before and after a given analysis with no untoward results noted, can eliminate malfunction as a possible source of error.

Calibration and maintenance records should be kept in the same manner as raw data, i.e., in bound notebooks, in ink, and suitably dated and signed. All results of calibration work should be noted. There is a temptation to ignore recording results of a calibration if everything seems normal, but it is important to record the fact that the calibration was done and results were as expected.

Maintenance work is frequently done by outside personnel, e.g., a manufacturer's representative or service technician. In this case the laboratory person responsible for checking the instrument after repair should document that the work was done, and by whom, and when. It should be stressed that *all* maintenance work, even minor items, e.g., the installation of a new light source or glass electrode, *must* be recorded.

Calibration and maintenance notebooks should be kept in the laboratory, in the vicinity of the instrument. If they are not; for example, if they are kept in the supervisor's office, there is always the danger of the information not being recorded. Each instrument should have its own notebook, for ease of reference, although the same notebook can be used for both calibration and maintenance.

ANALYTICAL REPORTS

The analytical report is the end-product of the analytical system. The analytical laboratory is an information-generating system and the report is the means of transmission of that information to the outside world. It follows that the information on the report should be complete, accurate, and easily understood by the client who submitted the sample. The following guidelines will assist in achieving these objectives:

1. The top of the first page, or introduction to the report, should include the following information:
   (a) The name of the laboratory which did the work.
   (b) The name and organization, if applicable, of the client who requested the work. Never give just the name of the organization without the name of the contact person—reports should be addressed to individuals, not organizations.
   (c) The date the sample was submitted or taken and the date the report was issued. In some instances, it may be important to give the time of day the sample was taken, or the time and date analysis was begun.
   (d) The origin of the sample—either submitted by the client or the name of the laboratory person who took the sample. Occasionally a blanket statement is used, e.g., "Unless otherwise designated, the sample was submitted by the client." While this is probably acceptable in the majority of situations, for full legal protection it is probably better to expressly designate the person who collected the sample or that the sample was submitted by a client.
2. The body of the report should contain the results of the analyses which were run, together with the following information:
   (a) The laboratory-assigned sample number.
   (b) The client-assigned sample number or other designation.
   (c) The name or appropriate designation of the analyte, i.e., the compound, element, or group of compounds for which the analysis was run. Care should be taken here to avoid ambiguities or possible misconceptions on the part of the client. Use chemical names wherever possible, but common names may be used if you are sure that there is no possibility of their being misconstrued. For example, the use of poly-(1,4-anhydroglucose) in place of the common name "cellulose" could only serve to confuse. On the other hand, abbreviations should always be avoided. The analyst may know that PCB stands for "polychlorinated biphenyl," but there is no guarantee that the client does.
   (d) The actual numerical result of the analysis. Care should be taken here. A result of "zero" should never be reported, since no chemist can state that there is zero concentration of anything in any sample. Instead, the "less than" sign, $<$,

should be used, followed by the detection limit of the analysis. The detection limit may have been experimentally verified, or may be estimated from a reagent blank. If the detection limit is unknown, the phrase "none detected" should be used, although this is less satisfying since it leaves open the question of how much could have been detected.

All analysts should be aware of the proper reporting of numerical results with respect to significant figures, i.e., that the digit furthest to the right should represent the only uncertain digit in the result. Non-degreed analysts (i.e., technicians) should be trained in the principles of reporting scientific data, or their reports carefully scrutinized. In particular, they are prone not to report trailing zeroes after a decimal point, when indeed the zeroes may be significant, or else go to the other extreme and report all the digits their electronic calculator can provide, regardless of significance.

(e) Some thought should be given to the reporting of units of concentration. In general, units should be chosen which clearly indicate whether the concentration is in terms of weight by weight, weight by volume, or volume by volume. Thus mg/L is preferred over ppm because the latter does not designate weight by volume. In dilute aqueous solutions, there is no significant difference, because the density of water is unity for all practical purposes, but in reporting the concentration of a solid contaminant in air, for example, there can be a large difference. In the latter case mg/M^3 would be a better choice of units. Concentrated aqueous solutions, often reported in percent, should be clearly designated as gm/100mL, g/100g, or % (by weight).

Some clients may prefer to have results reported in units which are not customary in analytical chemistry, e.g., g/ton or 1b/gal. In such cases it is wise to accommodate the client, since the analyst can be sure that the conversion is done correctly if he does it himself.

(f) Although not absolutely necessary, the author recommends that a code number be assigned to each analysis, consisting of the analytical method number and a number designating the analyst. This will be found to be very useful in tracing questioned data in the future.

(g) The actual format of the body of the report will depend on the nature of the report. For example, if a number of analyses were run on a single sample, the report may cover only that sample, with a table listing analytes, data, units and method/analyst code. On the other hand, the report may represent analyses for a single analyte on a multitude of samples of the same type, in which case the table will give sample number, analysis result, units and method/analyst code.

3. The bottom of the report should show the name, title and signature of the laboratory supervisor. This is a vital part of the report, since it is documentation that the supervisor has reviewed the results, approves them, and accepts responsibility for their quality.

PERSONNEL TRAINING RECORDS

Clients or users of analytical laboratory results assume that the data generated by the laboratory are the result of work by analysts thoroughly trained in laboratory operations. On the other hand, the question of adequate training may arise in cases of dispute over the accuracy of results. It is therefore good management practice for the laboratory supervisor to maintain files documenting the training of his personnel.

Professional (Degreed) Personnel

In general, the employee with a four-year college degree, with a major in chemistry, may be considered to have learned the basic laboratory techniques necessary to function as an analyst with minimum on-the-job training. He may not have had extensive training in modern instrumental analysis, but will have been exposed to the basic concepts of chromatography, absorption and emission spectrophotometry, and electrochemistry and should be able to master instrumental analysis using these techniques with a minimum of training and supervision.

Nevertheless, professional personnel interested in improving their scientific competency and skills will often turn to "continuing education" and take courses in special topics in chemistry, or in related topics, e.g. electronics, data processing, microbiology, or laboratory management. To the extent that such courses relate to the work of the laboratory, they should be recorded in the individual's personnel file, along with pertinent information, e.g. the institution offering the course, date completed, and whether or not the individual passed the course. Attendance at "mini-courses," seminars, and the like which may be given by instrument manufacturers or professional organizations should also be recorded in the personnel file. These records not only serve as documentation of relevant skills the individual has learned, but also are valuable reminders of the person's value to the organization at the time of the annual performance review. In fact, one excellent system of maintaining such records is to require each person to prepare a personal resume each year prior to performance review. Copies can be placed in the personnel file and used as evidence of continuing professional development. The preparation of the resume may also remind the individual who is "coasting" of the fact that he has done very little during the past year to improve his value to the organization.

Technician (Non-Degreed) Personnel

New technicians, unless they have previously been employed in a similar laboratory, are usually given on-the-job training by their supervisor, another professional, or an experienced technician. It is important that the extent of their training be documented, particularly as to the dates on which they were judged proficient enough to perform certain analyses. The dates are important because this will establish the extent of their experience at some future date when this might be questioned. The documentation should be related to their mastery of common laboratory operations, e.g., gravimetry, titrimetry, liquid/liquid extraction, or may simply indicate the date they were judged capable of performing a given analytical method. In the latter case, any relevant data should be recorded, e.g., the results of comparisons between the trainee and experienced staff on selected samples, or the results obtained on standard samples, e.g., Standard Reference Materials.

As in the case of professional employees, records should be kept of relevant courses taken, seminars attended, etc. Videotape training of technicians in selected topics is becoming widespread and will increase in the future because of the ability to train multiple personnel simultaneously, uniformity of training, and the advantages of the medium itself, e.g., the ease of upgrading the quality and content of the tape. The fact that a given technician has been exposed to a given videotape should be noted in his or her personnel file, along with the date and results of any test which may have been used to determine comprehension.

FILING QUALITY ASSURANCE DOCUMENTATION

The establishment of a good quality assurance program generates an amazing amount of paper, as the reader of this chapter will have guessed by now. To obtain the maximum benefit from this paper, a system should be established for rapid and easy retrieval of any given document. It should be remembered the quality assurance is undertaken for the benefit of outsiders, e.g., clients, government agencies, and accrediting bodies, and they will want to inspect the documentation of the system. In many of the laboratories the author has inspected, the request to see an example of a given QA document leads to a rapid scurrying a half a dozen offices before it is found. Occasions have arisen when it has not been found, even though there was no reason to believe that the paper in question did not exist. Embarrassment ensues for the laboratory manager.

The answer to the problem is a good filing system. All documents involved in quality assurance, or photocopies of them, should be placed in a central file, preferably kept in the Quality Assurance Director's office, or where he will have ready access. The actual system used may be tailored to suit the individual laboratory, and a good secretary can probably be invaluable in helping to define the system. The author claims no special expertise, but the following procedure is offered as a guideline for setting up a QA file.

First, it is necessary to decide which documents are going to be filed and to divide them up into relevant main groups. Each group is then given a title, and assigned a two-digit number. Within each main group, a series of file folders is established with a title and a number consisting of the main group number and a second two-digit number. For example, the main group number 02 may stand for Proficiency Evaluation Sample Reports, while a folder containing results reported to the Environmental Protection Agency may have the number 02–01, and the title "EPA—Potable Water." A master index is prepared, listing the various file folders by number, and each document to be filed has the appropriate number assigned to it prior to filing. In this way, improper filing is minimized. It is also good practice to only allow one person, the QA Director or a secretary to file or retrieve documents.

While each laboratory must determine its own filing system, the following is a possible list of main group titles:

Quality Control Charts - although these should be kept in the laboratory, copies should be placed in the file.
Proficiency Evaluation Test Results - file by the organization which issues the samples and the type of sample.
Standard Reference Materials - copies of certification documents and test results.
Blind Samples - file by type of sample.
Accreditation/Inspection Reports and Correspondence - file by organization.
Accreditation Certificates - file by organization.
Miscellaneous Quality Assurance Documents - copy of QA Manual, copies of standard operating procedures, QA audit documents, etc.

Copies of analytical methods for the QA Director's historical file are most conveniently kept in looseleaf notebooks. Instrument calibration data and equipment maintenance data are best kept in the laboratory where the notebooks used to record the information will be readily accessible to personnel. If kept elsewhere, the temptation is great not to record the data. For the same reason, personnel training records should be kept in the supervisor's office, with copies in the employee's personnel file, updated at least annually.

A good filing system will greatly increase the laboratory's efficiency and ability to respond to requests for information regarding quality assurance. It will also ensure that the relevant documentation will be retained for future evaluation.

7
COMPUTERS AND QUALITY ASSURANCE

When computers first burst on the technological scene after World War II, they were huge machines representing investments of millions of dollars, and conventional wisdom said that, while very useful for simple calculations on large volumes of data, e.g., government census reports, they would never find wide application in commerce and industry. This was, of course, before the invention of the transistor, and the power required to run and cool thousands of vacuum tubes was enough to discourage all except the government and the military from looking seriously at computers for practical applications.

However, the transistor *was* invented, followed by the revolution in solid-state electronics called "large scale integration" and "very large scale integration," so that today the thousands of vacuum tubes have been replaced by a few silicon integrated circuits or "chips," and the power consumption reduced to a few hundreds of watts. The computer revolution is in full swing as this is being written (mid-1984) and today, for about a thousand dollars, one can purchase a computer equivalent to the multimillion dollar machines of thirty years ago. There is no doubt that computers are here to stay, and we have now reached the point where the problem is not with the cost of the hardware, i.e., the machine itself, but with the software, or programs to run the machine.

Analytical chemists have watched the computer revolution progress with considerable fascination. Since analytical chemistry is concerned with the generation of information, mostly in the form of numerical data, it was obvious that, as hardware became less expensive and more available, applications in the analytical laboratory would abound. The first and most obvious applications came in the form of direct data acquisition from electronic instruments. After all, the output of these instruments is an electronic signal which was not difficult to couple, or "interface," to a computer, which could then manipulate the acquired data, producing the integrated area under a peak, for example.

These initial, pioneering attempts at direct data acquistion to a minicomputer were rather quickly superseded by the built-in microprocessor, in which the microcomputer is made as part of the instrument with direct LED data read-out, or printed "hard-copy" output. Microcomputers are so inexpensive today that they are being incorporated into virtually all instruments, including pH meters and balances.

Analytical chemists, being technically oriented, at first saw only the obvious data processing applications of computers. Other uses of computers, in areas such as information storage and retrieval, decision making, the control of information flow, were only slowly incorporated into the thinking of chemists and laboratory managers. However, in the last few years we have seen the introduction of various software systems into the laboratory which will have a profound effect on the future of laboratory management and operation. These have been developed and are being sold by numerous companies, some of them instrument manufacturers, and some of them software companies. They often are discussed under the acronym LIMS, for Laboratory Information Management Systems.

At the present state of the art, and in light of the very turbulent condition of the computer industry today, it would be presumptuous of any author to attempt to write a definitive chapter on computers with recommendations on hardware and software. The remainder of this chapter will be concerned with applications of the computer to enhance quality assurance, and other aspects of laboratory management, with which the author is familiar, i.e., it will describe aspects of computerization of the laboratory which *can be* and *are being done.*

SAMPLE HANDLING

"Accountability" of laboratory data has been described as those measures which assure that the data reported do in fact apply to the sample as it was submitted. This means that there has been no mix-up of samples, and the samples have been properly handled while awaiting analysis. One of the best ways of insuring accountability is by computer log-in.

A sample when received is entered into the computer, along with such pertinent information as the client's name, client's sample identification, and the analyses to be run, along with the analytical method number (for later calculation by the computer). The computer assigns a unique laboratory identification number to the sample and requests information on storage conditions and time of storage. The computer then assigns a storage location to the sample and prints out a label with all entered information, which is immediately attached to the sample container. The sample is then transported to the computer-assigned storage area to await analysis.

Note that, with a relatively simple program, the computer not only guards against sample mix-up, but also, through the use of "prompt" questions, forces the person entering the sample to think about such aspects of sample handling as storage conditions (i.e., refrigeration, freezing, inert gas and the like) and time of storage which are important factors in accountability of the final results. The immediate generation of a label which is affixed to the sample container guards against the common error of incorrectly applying labels when all labels are hand printed at the same time and then attached to the samples.

The information entered at sample log-in becomes part of a "data base" which can then be queried and manipulated for various purposes. For example, the computer can be asked to print out a list of all samples which have been analyzed and kept past their designated storage times. This will help prevent the buildup of a backlog of old samples which should be discarded. An analyst looking for work can receive a printout of all samples awaiting a particular analysis. The laboratory supervisor can receive a daily printout of all samples awaiting analysis in his laboratory and thus an immediate picture of his workload. Samples which are about to exceed a promised turn-around time can be given a priority rating. If fees are charged for analyses, as in a commericial testing laboratory, or an in-house analytical lab which cross-charges to other departments, a dollar value can be calculated by the computer for all "work in progress." There are, in short, innumerable ways in which such a data base may be used to improve laboratory management, other than quality assurance applications.

DATA ACQUISITION

By "data acquistion" is meant the means by which the raw data generated in the laboratory are entered into the mainframe computer used in the laboratory. There are, in general, two methods for putting data into the computer: direct acquisition and manual input.

Direct acquisition implies the use of electronic instruments which can be easily interfaced with the computer. Although it seems a logical way to go, there are many practical difficulties with direct acquisition. Many instruments, e.g., GC's, HPLC's and spectrophotometers require time spans of 20–30 minutes to generate a complete chromatogram or spectrum. It is usually not practical to expend this much computer time acquiring the data, especially since much of the data acquired is essentially meaningless. What we are interested in is the characteristic peaks in a chromatogram or spectrum, and not generally in the baseline data where there are no peaks. This consideration has led to the built-in instrument microprocessor which automatically calculates peak areas and peak positions. In theory, the ideal situation would be to dump this desired data (or selected portions of it) from the microprocessor directly to the mainframe computer, which could be done in milliseconds, but this requires interfacing the instrument microprocessor to the mainframe computer, which is often beyond the expertise of the analytical chemist.

What generally is done is that a human operator transfers the data from the instrument hard-copy readout to the mainframe computer. In other words, the data are manually transferred. Of course, this is also the input method used for all "wet chemical" or non-instrumental analyses. The major problem, of course, with all manually entered raw data is the possibility of human error. Transposition of digits, mix-ups of data and sample numbers, and incorrect entry are all possible when human beings handle numbers. Although the computer does not err like we humans, it also does not forgive, like the Divinity. Once the error is entered, it may be very difficult to detect unless the results are obviously wrong.

One system which is just beginning to be explored in some laboratories is the "electronic notebook." An electronic notebook is a portable, battery-operated computer about the size of a laboratory notebook which can be carried by an analyst around the laboratory. An analyst may sit at an electronic balance to weigh a series of samples. As each sample is weighed, he types the sample number into his electronic notebook and the weight is automatically entered through a direct link between the digitized output of the balance and the portable computer. Later, an automatic titrator may similarly communicate with the notebook, listing milliliters of titrant versus sample number. At some later time, the analyst dumps this data into the mainframe computer which stores it in memory and performs whatever calculations are necessary to obtain an analytical result.

QUALITY CONTROL DATA AND CALCULATIONS

In Chapter 3 we discussed techniques of using $\overline{X}$-charts and $\overline{R}$-charts for quality control purposes. In a large laboratory, running hundreds of different analyses, manual maintenance of such charts can become a monumental task. On the other hand, this is the type of work the computer does best: accumulating information, sorting it, processing it and displaying it in a variety of ways for human consumption. The following is one way in which this can be done.

For each analysis using check standards, a QC number is defined which automatically is entered in place of the sample number when the standard is run. This diverts the data to the QC program which accepts the raw data, calculates the result, and then updates the mean and standard deviation which have been calculated from previous standards. The date of analysis and employee number of the analyst are also retained by the computer.

On a periodic basis, usually monthly, a QC printout is generated, one copy of which goes to the Quality Assurance Director, and another copy to the laboratory supervisor. An example of such a printout is given in Figure 7–1.

Note that the heading of the report gives the QC number for this particular analysis. The material used for reference is also designated as well as the analyte and the particular analytical method number. The designation "certified value" will be followed by the known concentration of the analyte, if this is known. This is followed by the cumulative mean and the 95% confidence range, estimated by twice the standard deviation. To be precise, the Student *t*-factor should be used, but this would involve storing a table of Student *t*'s in computer memory and was judged an unnecessary refinement. This is followed by the number of results which have been outside the two standard deviation limits for the year to date (YTD), and the date of the last result outside the limit.

The program is designed so that, if a result exceeds these limits, when the data is entered a message is flashed on the computer CRT

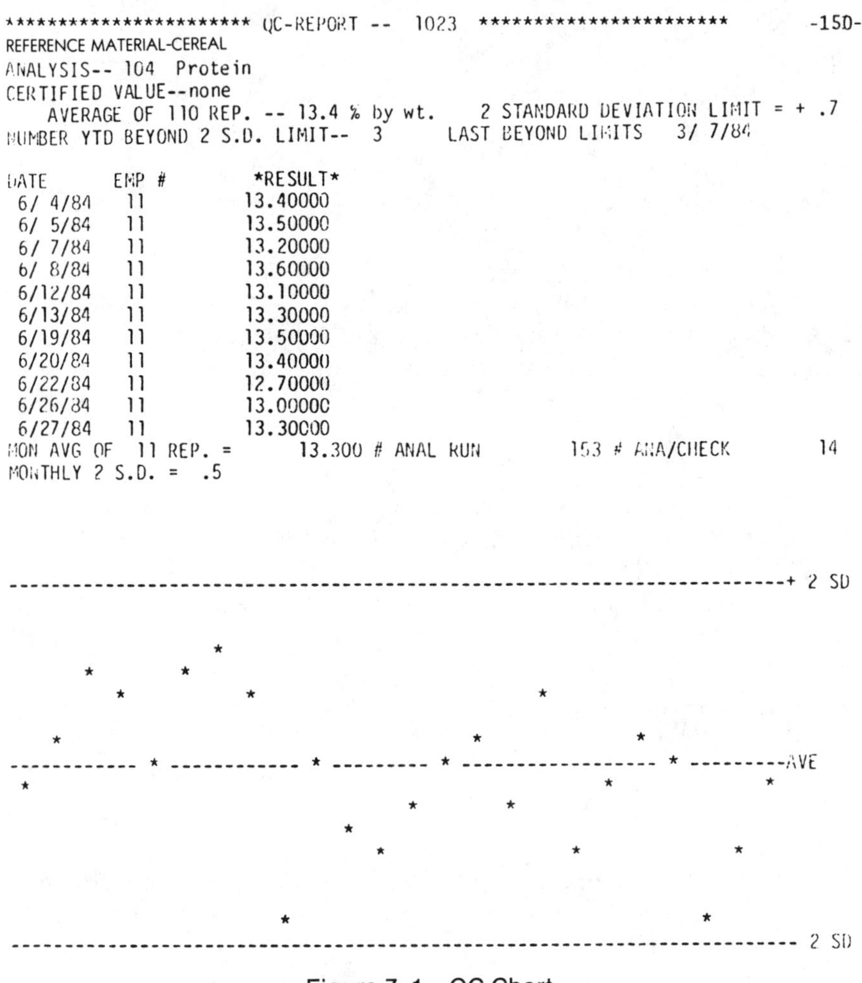

```
*********************** QC-REPORT -- 1023 ***********************          -15D-
REFERENCE MATERIAL-CEREAL
ANALYSIS-- 104 Protein
CERTIFIED VALUE--none
   AVERAGE OF 110 REP. -- 13.4 % by wt.      2 STANDARD DEVIATION LIMIT = + .7
NUMBER YTD BEYOND 2 S.D. LIMIT-- 3        LAST BEYOND LIMITS    3/ 7/84

DATE        EMP #         *RESULT*
 6/ 4/84     11           13.40000
 6/ 5/84     11           13.50000
 6/ 7/84     11           13.20000
 6/ 8/84     11           13.60000
 6/12/84     11           13.10000
 6/13/84     11           13.30000
 6/19/84     11           13.50000
 6/20/84     11           13.40000
 6/22/84     11           12.70000
 6/26/84     11           13.00000
 6/27/84     11           13.30000
MON AVG OF   11 REP. =       13.300 # ANAL RUN         153 # ANA/CHECK        14
MONTHLY 2 S.D. =   .5
```


Figure 7–1 QC-Chart

screen which so informs the analyst and instructs him/her to see his/her supervisor immediately. This assures that the supervisor is alerted and that action is taken, even if only a second run of the standard to confirm whether or not the system is out of control. Of importance also is the fact that once the data has been entered, it cannot be deleted or changed by the analyst.

To return to Figure 7–1, the three columns give the date, employee number, and result obtained on the standards run during the month.

Beneath these data are the monthly average and two standard deviation limits, the number of analyses run during the month, and the ratio of analyses to check standards.

Finally, the computer plots the $\overline{X}$-chart for the last twenty-five check standards run. If necessary, a chart could be generated for all of the standards run since the beginning of the program, but this is usually not practical or desirable. By examining the chart, the Quality Assurance Director or the laboratory supervisor, using the techniques described in Chapter 3, can quickly determine any tendencies to drift out of control and take appropriate action.

Similar charts can be used to display data from spiked sample analyses, except that in this case the percent recovery of spike is calculated and plotted, rather than the concentration of analyte.

Duplicate analyses can be handled the same way, as illustrated in Figure 7–2. Here, the range of duplicates is plotted, along with the 50% and 95% confidence limit lines. In the illustration, the "percent range," i.e., the range divided by the average of the duplicates, is shown rather than the range itself.

COMPUTER GENERATED ANALYTICAL REPORTS

If a computerized sample log-in system is used along with a numbered analytical method system, as described in Chapter 6, it is a relatively simple task to program the computer to perform all necessary calculations to obtain final results. The analyst manually enters the raw data in conjunction with the sample number and the analytical method number in response to a series of computer "prompts" and the calculated result is displayed at the computer terminal. When all analyses have been completed, the computer may be programmed to generate a printed report.

However, before final report generation, it is desirable that some degree of control be exercised by the laboratory supervisor. To accomplish this, a status symbol or flag is attached to each analysis for each sample. If the analysis has not been run, the status may be indicated by the letter "I" for "incomplete." Once the raw data is entered and calculations performed, the status changes to "C" for "completed." However, no report will be generated until the supervisor reviews results and changes the status of all analyses to "V" for verified.

6/29/84
QC DUPLICATE ANALYSIS SYSTEM

220 Nitrate Nitrogen		# ANA	# DUP	C.V.	#ANA/DUP	AVG R
	MTD	191	7	2.1528	27.28	2.43
	YTD	813	84	2.0608	9.67	2.32

DATE LAST OUT 6/28/84 # OUT 27 LIMIT OF SENS 0.02000 mg/l

SAMPLE NO.	DATE	ANALYZED BY	RESULT#1	RESULT#2	R VALUE	
306815	6/ 8/84	103	2.2200	2.2600	1.7857	0.0400
307214	6/ 8/84	103	7.4700	7.6300	2.1192	0.1600
307919	6/ 8/84	103	16.5800	16.6000	0.1205	0.0200
310531	6/21/84	103	1.4900	1.4800	0.6734	0.0100
310663	6/21/84	103	11.7400	11.4600	2.4137	0.2800
311444	6/28/84	103	9.6900	8.8900	8.6114	0.8000
312138	6/28/84	103	3.8600	3.8100	1.3037	0.0500

```
**** CHART OF R VALUES ****
   6.05            #                        #               #
   5.82
--------------------------------------------------------------
   5.58
   5.35
   5.12
   4.89
   4.65
   4.42
   4.19
   3.95
   3.72
   3.49
   3.26
   3.02                                  *  *  *
   2.79
   2.56                            *
   2.32                                                      *
   2.09                                                *
--------------------------------------------------------------
   1.86                                              *
   1.63                                   *
   1.39                                                         *
   1.16             *      *        *
   0.93
   0.69                                                    *
   0.46
   0.23
   0.00        -*-*----*-*---*-*--------------*-*-----*----------------
```

Figure 7-2 *R*-Chart

The advantages of this type of computerization for quality assurance are twofold. First, calculation of results by the computer eliminates calculation errors which are all too frequent when human beings do them, even with electronic calculators. Second, the requirement of supervisor verification before report generation ensures that at least one experienced, technically trained individual has examined the data before it leaves the laboratory.

SECURITY CONSIDERATIONS

As computers begin to control and organize laboratory operations, consideration needs to be given to security with respect to safeguarding against malfunctioning of the computer, loss of data, and unauthorized access to data in the computer.

Before any program is used in real-time operations, it should be thoroughly tested, using dummy data, and comparing results with expected results or hand-calculated results. Once properly programmed, the computer can be relied upon to be error-free, barring electrical breakdown. However, many complex programs which are designed to handle a wide variety of situations may have unexpected "glitches" which only show up under specific conditions of input/output. If possible, a new program should be tried out in one department or a small section of operations before being adopted for use by the entire laboratory.

Computers are electronic devices and depend on a steady supply of electricity. Unfortunately, in many areas of the country, the supply of mains electricity can be and is interrupted by thunderstorms, ice storms, etc. When this happens, of course, the computer is useless until power is restored. What is worse, all data in "random access memory" (RAM) is lost as well. This means that data in the computer's memory should be transferred to magnetic disk storage as soon as possible. Even here the data is not secure, since disk drives do occasionally suffer "head crashes" which tend to wipe out all data on the disk. Therefore, the disk currently being used should be backed-up onto another disk at frequent intervals, i.e., at least once or twice each day. In this way, only a half-day or full-day's data will be lost. If the analytical data are manually recorded or in the form of instrument generated "hard-copy," it will have to be reentered, but this is better than having to rerun all analyses. Back-up disks should be stored in a remote location as a further security provision.

One ever present potential hazard in electrically operated equipment is fire. Fires have occurred in data processing centers which have not only destroyed days of carefully accumulated data, but the computer and its peripherals as well. However, in many cases more damage was done by the means used to put the fire out, i.e., water or dry powder fire extinguishers, than by the fire itself. Therefore, it is good practice to equip the computer center with one or more *halon*-type fire extinguishers which leave no residue after they are used.

Access to data in the computer should be restricted to personnel on a "need to know" basis. For example, only certain persons may be permitted to log-in samples, while others may be permitted to enter data from analyses. Only supervisors or their suitably designated delegates should be allowed to verify analytical results. Quality control data should be restricted to supervisors, the Quality Assurance Director, and senior management.

This restricted access is accomplished by assigning to each person who uses the computer a password or key, i.e., a unique number or letter/number combination and programming the computer with regard to the various programs available to that password. Obviously, this will not prevent unscrupulous persons from obtaining and using someone else's password, but it will prevent accidental or casual tampering with the system. If the situation requires tight control of access, management can deal with it by express directives concerning the confidentiality of passwords and spelling out stringent penalties for improper use of another's password.

HARDWARE AND SOFTWARE

As mentioned in the introduction to this chapter, in the current state of the computer industry, it is impossible to give firm recommendations regarding actual hardware and software which will perform satisfactorily in today's analytical chemistry laboratory. Many factors need to be considered before purchasing computer power: the size of the laboratory, the nature of the work being done in the laboratory, the amount of funds available for computerization, and the existing skill level with regard to computer programming. Nevertheless, some general comments may be in order, bearing in mind the current chaos in the computer world and a recent comment by a computer expert to the author that "any computer you buy today will be obsolete in three years!"

Of prime importance in considering computerization of the laboratory is the current and projected scope of activities which the computer may be called upon to do. The applications discussed in this chapter are only a small part of the capabilities of today's computers. For example, word processing on a computer can greatly enhance the productivity of a secretary or secretarial staff. Various lists or data bases may be found useful. A list of instruments, with manufacturer, model number, date of purchase, price, etc. will be useful for inventory and insurance purposes. A list of clients or users with information on number of samples and type of analyses can be very helpful for budgeting and long-term manpower projections. Employee time records and other cost information may assist in evaluating costs of analyses and setting fees. Remember that the computer is somewhat like the camel with his nose in the tent: once employees get used to the first computer, numerous possible applications spring to mind and more computer power is required! However, one small caveat—remember always that just because something *can* be done on a computer doesn't mean that it *must* be done or that it is necessarily desirable to do it on the computer. For example, storage of data which is seldom used in computer memory may well be a waste of computer memory. Pieces of paper and filing cabinets have not yet been totally displaced by magnetic disks.

Computerizing a laboratory from scratch is a very difficult problem because we are dealing with the interface of two scientific disciplines: computer science and analytical chemistry. Unfortunately, few analytical chemists have expertise in computer programming, and even fewer computer programmers are familiar with the work of an analytical chemistry laboratory. While instrument manufacturers and software companies are addressing this problem, their results are often less than totally satisfactory because of the wide variety and diversity of analytical labs which seem to demand almost custom design of the computer system. In general, it is not practical or sometimes even possible to reorganize the laboratory to fit the computer system. On the other hand, the multiple benefits of computerization are a powerful stimulus toward making full use of this powerful tool.

With the current popularity and relatively low cost of microcomputers, or "personal computers," it is tempting to begin with the purchase of one of the higher quality brands and gradually build up computer power. Except for the very smallest of laboratories, this is not a very good idea. One microcomputer can be used by only one person at a time and it will soon become apparent that computer time is a bottleneck. Of course, more microcomputers can be purchased, which will alleviate the situation temporarily. However, multiple microcomputers will require multiple programming, duplication of data bases, etc. unless they can be linked together, which is often not possible. A much more satisfactory system is a minicomputer or central processing unit which can accept input from, and generate output to, multiple terminals located throughout the laboratory operating on a time-sharing basis. These terminals can then share common data bases and transfer information to and from the central unit or each other.

In addition to the basic computing capacity, various peripherals will be required. Data and programs must be stored, and this requires one or more disk drives. Floppy disks or diskettes will only be useful in the very small laboratory, while most labs will require the larger (and more expensive) hard disk drives.

One or more printers will be necessary to retrieve data from the computer in hard-copy format. If the computer is to be used for word processing, a letter quality, daisy wheel printer may be very desirable. A telephone modem will also be required if data is to be transferred to remote locations and may also be found useful for tapping into one or more of the large data bases of chemical information which are currently being maintained by various organizations.

Training of personnel to use computers is much less of a problem today than it was a few years ago due to the proliferation of personal computers, videogames, and high school and college computer courses. Nevertheless, there is still a significant proportion of the population which is fearful of using a computer. These fears are usually of two kinds: they fear that the machine is too complicated to be mastered, or they fear that something they do will "break" the computer. The only way these fears can be overcome is by hands-on experience at a terminal. One of the best ways of accomplishing this is through the use of one or more of the games that are available in all computer software packages. Permission to use the computer for game playing during lunch hour or after work will easily familiarize the timid employee with the machine and overcome his/her fears.

There can be no doubt that the computer is here to stay. Electronic instrumentation has done much to improve the productivity of the analytical laboratory as an information source, and the infant science of laboratory robotics is just beginning to show its potential in improving productivity. The role of the computer in the laboratory of the future will be to control, monitor, and display the resulting information.

8
ESTABLISHING A QUALITY ASSURANCE PROGRAM

The previous chapters have described the elements of a comprehensive quality assurance program. This chapter is concerned with the problem of how to establish such a program in an existing laboratory which may have some of the required elements, but not all. In the author's experience, many laboratories have various aspects of quality control in place, but are deficient in quality assurance. Quality control is essentially good scientific practice, but the elements of documentation; sample control, traceability, etc., are often woefully lacking in many modern laboratories.

MANAGEMENT COMMITMENT

It is absolutely essential that top management in the organization be committed to the establishment of a comprehensive quality assurance program, and that this commitment is communicated to all levels of personnel. In addition to management directives, memos, and oral statements, management must back up its commitment by action, specifically by demonstrating a willingness to commit the necessary financial resources for quality assurance, designating authority and responsibility for quality assurance, and holding technical personnel accountable for their quality assurance responsibilities.

Before management makes this commitment, however, it is necessary that managers themselves be convinced that their commitment is both necessary and desirable. Quality assurance is a cost item which can be readily and easily identified by standard accounting procedures. The economic benefits on the other hand are not as easily identified, but are nevertheless real. Quality assurance costs may be considered in the same light as safety programs, security, property insurance, etc., i.e., as elements of a risk management or loss prevention program. The value of a quality assurance program depends to a certain extent on the uses to which the information generated in the laboratory is put, and this will depend on the laboratory's place in the overall organization, or in the commercial world. The following will illustrate some of these considerations.

Data generated in the laboratory may be used as the basis for the expenditure of large sums of money. Consider the typical project involving control of effluent waste from a manufacturing plant, whether dictated by government regulations or by management's social conscience. To assess the extent of the problem, samples are taken and analyzed to establish the existing situation. After analysis, engineers are engaged to design, fabricate or purchase pollution abatement equipment. Pilot plants may be built and operated, requiring more samples and analyses. Finally, when the abatement equipment is in place the new system must be evaluated for the extent to which objectives have been met. The point is that data generated by the lab are critical to the entire process, and therefore the quality of that data and its reliability are critical.

The cost of the information provided by the analytical laboratory is escalating rapidly, primarily due to the need for sophisticated instrumentation to achieve the limits of sensitivity required in today's world. Modern instruments, e.g., the gas chromatograph/mass spectrometer, or inductively coupled argon plasma spectrophotometer cost hundreds of thousands of dollars. In addition, highly trained personnel are needed to run them and interpret the results. Thus, the cost of analyses per sample is in the hundreds to thousands of dollars. In general, as the cost of any information increases so does concern about its reliability or quality. The costs of a quality assurance program are a relatively small price to pay for data of known reliability.

Many laboratories today are facing the need for accreditation. (See Chapter 9.) Accreditation is the certification by an outside organization that the laboratory is capable of generating dependable data. All accrediting agencies today demand that the laboratory have a comprehensive quality assurance program. Lacking such a program, the laboratory will be denied accreditation which can mean loss of revenue from government projects, less credibility in dealing with customers or suppliers, and greater difficulty in defending analytical data in forensic situations.

The extent to which a given laboratory finds itself involved in situations such as these will define the need for a quality assurance program and the comprehensiveness of the program. Top management must decide the need and extent of commitment to quality assurance.

THE QUALITY ASSURANCE DIRECTOR

Personal Characteristics

The title "Quality Assurance Director" is used throughout this book to designate the person with the authority and responsibility for assuring quality in the laboratory. Many other titles are used in industrial organizations, e.g., Quality Assurance Manager, Supervisor, Coordinator, etc., and the choice of title may depend on personal preference or conformity to whatever system is used in the organization. In manufacturing operations, with existing organizations concerned with product quality control, the title may be preceded by the "Laboratory," as in "Laboratory Quality Assurance Director," to emphasize the fact that the responsibilities are not concerned with product quality, but the quality of laboratory data.

In large laboratories it may be necessary to have a staff of personnel devoted to quality assurance, while in a small laboratory there may not be enough quality assurance work to engage a full-time Quality Assurance Director. The size of the quality assurance staff will depend on the volume of analyses run and to some extent on the variety of analyses. A large laboratory running many similar samples for a limited number of analyses will require less quality assurance documentation, data, etc. than a somewhat smaller laboratory which may be running a wide variety of analyses on a broad spectrum of samples. As a general rule, in the author's experience, a full-time Quality Assurance Director is not needed until the technical staff begins to approach about one hundred persons.

The choice of a suitable person for Quality Assurance Director should be carefully considered, with the following criteria in mind.

Education and Experience. The person chosen should have a minimum of a bachelor's degree in chemistry, with some years of experience in analytical chemistry. In the author's opinion, the requirement of a B.S. in chemistry is essential. The Quality Assurance Director must interface with both the laboratory's clients and technical personnel, and must be able to call on a chemical background to understand both constituencies. While biologists, chemical engineers, medical technicians, etc. may have had some training in chemistry, they do not have the depth of knowledge necessary to function effectively without calling for chemical assistance, which weakens their position.

It is also necessary for the Director to have experience in analytical chemistry to understand the problems faced by laboratory personnel and to assist in their resolution. A chemist whose career has been spent synthesizing organic compounds cannot be expected to easily acquire the thought habits or mind-set of an analytical chemist.

Knowledge of Statistics. As demonstrated in previous chapters, a familiarity with statistical evaluation of data is a basic requirement of a person charged with responsibility for quality assurance. A degree in mathematics is not necessary, but the QA Director should be comfortable with the methods of statistical analysis and be able to interpret the results of such analysis. Statistical analysis is not a difficult task, especially in today's world of computers, but the concepts of statistics and its parent field of probability can be rather subtle and easily misinterpreted. While the practical aspects of statistical analysis can be obtained from reading the literature, it is recommended that the QA Director have taken at least one good course in mathematical statistics as part of his/her background.

General Characteristics. The QA Director should be a person of maturity and be able to interact effectively with other individuals. The duties of the position will require contact with laboratory personnel at all levels as well as with outside persons representing clients, accrediting agencies, government personnel, etc. In dealing with laboratory personnel, the QA Director must represent the clients and their concern with quality data. He/she must be prepared to shut down a portion of the laboratory if results are not up to his/her standards of quality. On the other hand, he/she must be prepared to discuss and demonstrate the quality assurance program to outsiders. Since a good part of the job will involve writing letters and reports, mastery of the English language is essential. It would also be desirable if he were "computer literate," at least to the extent of being able to communicate with a computer and understand elementary computer programming.

The Quality Assurance Director's Place in the Organization

There are two important criteria involved in placing the Quality Assurance Director in the organizational structure. The first of these is that he/she must report directly to top management, and the second is that he/she must not have direct responsibility for the generation of the analytical data.

The Quality Assurance Director should either be a member of top management or responsible solely to top management. In no circumstance should he/she report to the laboratory supervisor, technical director, or other person responsible for technical supervision. The reason for this is that he/she must be perceived by both laboratory personnel and persons outside the organization as an independent monitor of the quality of the data being generated in the laboratory. In other words, there should be no real or perceived conflict of interest.

This requirement of independence can be a difficult one to satisfy in many laboratories. If the laboratory is not large enough to require a full-time Quality Assurance Director, it is unlikely to have a person available with the necessary technical background who is not also involved in generating analytical results. The following are some ways of circumventing this problem.

If the laboratory is part of a large organization which employs large numbers of chemists, it may be possible to appoint a chemist not working in the analytical laboratory to the position of Laboratory Quality Assurance Director, with the understanding that he/she will devote a certain percentage of his/her time to this task. Such a person might be borrowed from the R & D department, library, engineering department, or technical service department, for example. Needless to say, he/she should be working in an area geographically close to the analytical laboratory. It is virtually impossible to perform effectively as Quality Assurance Director from a remote location by mail or telephone.

Large organizations often have more than one analytical laboratory located at the same site. In this case, an ideal solution of the problem is to exchange personnel as Quality Assurance Director, i.e., a chemist in lab A can serve as QA Director for lab B, and vice versa. Such an arrangement can offer excellent training for the chemists involved, will improve communications between laboratories, and improve laboratory operations by cross-fertilization of ideas.

A variation on this can be applied in the case of a single laboratory, provided it is large enough to be divided into several departments. An individual can be appointed to the position of Quality Assurance Director for all departments except his/her own, and an individual from one of the other departments is appointed to monitor the work in the QA Director's primary technical responsibility. Both individuals report separately to top management.

A final possibility which may be explored, if none of the above situations can be implemented, is to hire a part-time Quality Assurance Director. Many communities contain retired analytical chemists, housewives and mothers with analytical chemistry backgrounds, and college professors and high school teachers who would be delighted with the opportunity to work part-time and simultaneously maintain contact with their profession. A retired chemist from the laboratory itself would be an ideal choice for such a position, as he/she is already thoroughly familiar with the work of the laboratory.

DEFINING THE QUALITY ASSURANCE PROGRAM

Once the Quality Assurance Director has been chosen, the next step is to define the details of the program to be established. The Director, the laboratory supervisor, and the supervisor and Director's immediate superior should meet to decide the need for quality assurance and the extent of detail required to achieve it. This will depend on the uses to which the lab's data are put, as described previously, and the principal objectives of quality assurance, as described in Chapter 1. Questions such as the following should be asked and answered:

- What are the consequences of poor quality data, i.e., what situations have arisen in the past which could have been avoided if a good QA program had been in place?
- How often does the lab repeat analyses because clients refuse to accept data when first presented?

- Does the lab need a QA program for accreditation purposes, either now or in the immediate future?
- Are there important company programs currently in place, or contemplated in the future, which will require good quality data to avoid disaster?
- What quality control procedures are currently being followed? Are instruments being calibrated frequently, standards being run, and training programs used for new personnel?
- What quality assurance practices are being followed? Is there any protection against sample mix-up? Are samples being logged? Are analytical methods well defined and authorized? Can results be traced if they are questioned?

When these and similar questions have been discussed, the QA Director should be able to put together the outline of a quality assurance program based on the needs of the laboratory and those areas of vulnerability to criticism covered in the discussion. At this point, it is time to bring into the discussion the technical personnel in the laboratory. A meeting should be held with at least key technical personnel, preferably with all technical personnel. The meeting should be chaired by a representative of top management who should declare the organization's commitment to improved quality assurance and the delegation of responsibility for quality assurance. The QA Director should then present the objectives of the program (i.e., QC documentation, traceability, accountability, security of data) and his outline of the program as he currently envisions it. The points should be made that all technical personnel will have the opportunity to critique the final program, and that every attempt will be made at consensus, i.e., that agreement will be reached regarding necessary operations to achieve goals.

WRITING STANDARD OPERATING PROCEDURES

At this point, based on the previous discussions the QA Director may begin drafting the written form of the program. A written document is necessary for two reasons. First, a written document is less subject to misinterpretation, confusion, later changes in defining what was meant or agreed upon than oral instructions; and second, writing a program forces in-depth thinking about the content of the program.

In the author's opinion, the best format for the written Quality Assurance Program is as a series of Standard Operating Procedures, or SOP's. A Standard Operating Procedure is a document which instructs someone how to perform a task. It is essentially a no-nonsense "how to do it" document which establishes the approved or required procedure for accomplishing an objective. The following is a recommended format for written SOP's.

Number. Every SOP should be given a number so that it may be unambiguously referred to in other documents. If an organization uses SOP's for a multitude of purposes, an alphanumeric designation may be used, which can indicate the general area of the subject, e.g., "QA-1" would be the first SOP concerned with Quality Assurance.

Title. The title should be brief but descriptive. Remember that the title is the means by which a searcher would look for an SOP in an index, so the title should give enough information to enable him/her to find the relevant SOP.

Background Information. This section may or may not be included depending on how the author assesses the needs of the reader with regard to the particular information in the SOP.

Scope. This section should delineate the field of application of the SOP to avoid ambiguity over whether the SOP should be universally applied, only in certain specific instances, etc.

Purpose. SOP's can be written to cover virtually every task undertaken in a laboratory or any other organization. In most cases they would be totally unnecessary. If a good reason or purpose cannot be described for an SOP, it should not be written.

Operations. This is the heart of the SOP. It should consist of a series of numbered paragraphs which take the reader through the actions necessary to achieve the purpose of the SOP. Care should be taken that the paragraphs are written as clearly and as free of the possibility of misinterpretation as possible. Avoid the use of words "should," "could," or "may" since these give the reader an option.

Instead use the words "shall" or "will" whenever possible. Remember that the SOP is a series of instructions which must be followed, not a set of guidelines which the reader may follow or not as he chooses.

SOP's may be long and involved or relatively short. Long, involved SOP's should be scrutinized to see if the purpose might be better served by breaking them into several short ones. Figure 8-1 is an example of a short SOP covering one topic of a Quality Assurance Program.

TOPICS FOR STANDARD OPERATING PROCEDURES

At this point, the reader may be wondering where to start in preparing a set of SOP's for his/her particular laboratory. As mentioned above, the actual program will have to be custom-designed to fit the laboratory involved. However, the following is offered as a list of topics which should be considered. Details and recommended procedures will be found in other relevant chapters of this book.

Laboratory Notebooks. How issued, to whom, where stored when completed, how kept (ink, dated, signed, etc.)

Analytical Methods. Maintenance of analytical methods manuals, format for written analytical methods, system for numbering methods.

Sampling Methods. Maintenance of written sampling methods manual.

Criteria for Authorized Methods. How methods are selected, validated, authorized, and necessary documentation.

Equipment Calibration. Equipment calibration schedule, methods of calibration, documentation of calibration.

Equipment Maintenance. Necessary preventive maintenance, documentation of maintenance.

Reagents. Selection of reagents, labeling of reagents with date received, opened, and expiration date.

QUALITY ASSURANCE OPERATIONS MANUAL

Standard Operating Procedure QA-4

Title: Reagents and Reagent Solutions

Scope: This operating procedure will cover ways and means of preparation and storage of reagents and reagent solutions in the laboratory.

Purpose: The purpose of this procedure is to insure good quality control in laboratory operations, and to permit traceability of possible causes of error in analytical results.

Definitions: A *reagent* is defined as any chemical used in a chemical analysis or microbiological test, other than the sample being analyzed.

A *reagent solution* is a solution or other mixture of chemicals prepared in the laboratory for use in a chemical analysis or microbiological test.

Procedures:

(1) All reagent chemicals received in the laboratory will be labeled with the date of receipt by the Purchasing Agent. A date when the bottle was opened and an expiration date will also be placed on the label by the analyst authorizing the purchase, if applicable. The only exception to this rule will be reagents used in high volume, e.g., extraction solvents, which are used up in a relatively short period of time.

The dates will be recorded on a separate label placed on the bottle, and the label placed so that it does not obscure the manufacture's label. Suitable labels are commercially available in tape form. The common practice of scratching the dates on the manufacturer's label is not good scientific practice and is not acceptable.

(2) At the time the reagent bottle is labeled, an adhesive-backed colored dot corresponding to our safety code shall also be placed on the bottle. This will be done by the Purchasing Department in consultations with the manager of the program responsible for the purchase. The safety color code is as follows:

Green	-	non-toxic
Red	-	flammable or combustible
Orange	-	moderately toxic (LD_{50} - 200 to 1000 mg/kg)
Double Orange	-	very toxic (LD_{50} - 50 to 200 mg/kg)
Blue	-	toxicity unknown or undertermined

(3) All reagent solutions when prepared will be labeled with the date of prepartion and an expiration date. If shelf life is known or specified in an analytical method, the expiration date will correspond to the shelf life. If the shelf life is not known (i.e., the solution is assumed to be stable), an expiration date of one year from the date of preparation may be assumed unless a longer stability time is documentable. The only exceptions to this rule will be solutions which are prepared fresh for each analysis or batch of analyses and discarded when analyses are completed.

(4) All data pertinent to reagent preparation and standardization will be logged in a separate, bound notebook kept in the laboratory or in the laboratory notebook of the analyst who prepared the solution. Each entry in the notebook should be dated and signed by the person who did the preparation or standardization.

Figure 8-1 Standard Operating Procedure for Solutions

Reagent Solutions. Documentation of solution preparation, labeling of solutions.

Sample Handling. How to log samples, identification of samples, storage of samples, disposal of samples.

Check Standards and Duplicate Analyses. How often run, charting of results, reporting of results.

Reporting of QA Deficiencies. Mechanism whereby QA Director reports QA problems to top management.

Personnel Training Records. Documentation of personnel training, where training records are kept.

Quality Assurance Files. File system used, where files are kept, who has access to files.

CONSOLIDATING THE PROGRAM

Once the quality assurance Standard Operating Procedures are written, they should be circulated in draft form to key technical personnel, i.e., laboratory supervisors, senior chemists, or others with authority and responsibility for the technical work of the laboratory. After sufficient time has elapsed for these personnel to have read and digested the SOP's, a second meeting should be called for an in-depth discussion of the program, based on the written SOP's.

At this meeting, the point should be made and stressed that what is not wanted is a ''paper'' program, i.e., a program which exists only in the written SOP's, but does not correspond to the reality of actual laboratory operations. What is sought is an agreement that the SOP's are workable, and that the laboratory personnel will put them into practice. In general, two types of objections will be raised. The first will be a complaint that one or more SOP's are simply not practical for one reason or another. In this case, the Quality Assurance Director should make every effort to investigate whether the complaint is justified and, if so, to eliminate the SOP, or possibly insist on a change in operations to accommodate the SOP, if the laboratory will be at a serious risk if it is not incorporated. A third alternative is to compromise to achieve the end desired by a different means.

The second type of complaint will be essentially trivial, based on the common feeling of "We have never had to do things this way in the past. Why do we have to change now?" Although the Quality Assurance Director may see the triviality of the complaint, it is not wise to dismiss this type of complaint out of hand. A good Quality Assurance program is going to require that people change the way things are done, and it is human nature to resist change without good reason. The QA Director should explain the basic reason for the change in terms of the objectives of the program and the vulnerability of the laboratory if the change is not made.

It may be tempting for the QA Director to exercise his authority at one point or another and simply state that this is the way things are going to be done. This is not wise if it can be avoided. The point of this whole exercise is to achieve agreement on the part of technical personnel so that they will put into practice the SOP's as they are written. If the program is imposed on them from above, there will be a temptation to ignore parts of it, or to attempt to foil the system to prove their point that it is unworkable. It may be necessary to hold several meetings with interim revisions of the SOP's before total consensus on the program and agreement to abide by it are achieved.

MONITORING THE PROGRAM

Once agreement has been obtained, a period of a month or two should be allowed to put the program into effect. The QA Director should be available during this time for consultation and discussion regarding any problems which may arise. Once again, compromise and ingenuity may be required (together with revision of SOP's). An example will illustrate this.

One department of a laboratory in which the author was employed objected to the requirement that analysts sign their notebooks with full, legal signature, insisting that initials were sufficient. The QA Director was inclined to agree until he realized that there were at least two employees of the laboratory (not in the same department) with the same set of initials. One compromise that was suggested was to use employee number, rather than initials. This would have been acceptable, except that in this company employee numbers were often "recycled" to new employees who might be given a number which had once belonged to an employee who had left. The compromise which was finally arrived at was to use both employee number and initials, since it was highly improbable that two employees would both have the same initials and employee number. Needless to say, in a forensic laboratory or a laboratory whose results are often required to undergo legal scrutiny, the full, legal signature should always be used.

After sufficient time has elapsed, the QA Director should conduct a full QA audit of the laboratory, department by department, to assess compliance with the SOP's. A full report should be issued to top management reporting the degree of compliance and any problems which are uncovered.

With the program in place and operating, the QA Director can confine his attention to monitoring the system and collection of data to be used in support of replies to problems involving the quality of the data generated.

Quality Assurance Audits

The Quality Assurance Director should periodically audit the program to ensure compliance. There are two ways in which this may be accomplished. One is to conduct a full-scale audit, complete with checklist, of each department, covering all applicable SOP's, in conjunction with the laboratory supervisor. The other method is to audit each department on one aspect of the QA program, on a random basis, preferably discussing with selected technical employees the extent to which they are complying with whichever SOP is under audit. Ideally, the two methods should be used. For example, the full-scale audit may be undertaken on an annual basis, while the random audit may be conducted as time permits. Just as documentation is the key to quality assurance operations in the laboratory, so documentation of the QA audits by the QA Director is necessary. This is usually in the form of a written report to top management.

Monitoring Quality Assurance Data

The QA Director should become the collector and focus of all data generated in support of the QA program. This will include periodic (monthly at least) copies of QC charts, results of proficiency sample tests, results on blind samples, copies of all correspondence related to quality assurance and accreditation, new methods authorization and validation data, etc. He/she will also keep comprehensive files of all information related to quality assurance.

The blind sample program is the best probe of the quality assurance program. This program should be directly administered by the QA Director, with all results directed to him/her, and QC charts kept by him/her.

Reporting Quality Assurance Problems

Since the QA Director receives his authority from, and is responsible to, senior management, it is his/her duty to report to his/her superiors any problems or irregularities detected in the quality system, especially since the ultimate legal responsibility for the quality of the data generated rests with top management. The dictum "Management does not like surprises" applies here.

To ensure that the QA Director does not neglect this aspect of his job, a mechanism should be established for reporting quality assurance irregularities to top management. One way of doing this is through the use of a form, as illustrated in Figure 8-2. The term "irregularity" is used in place of "deficiency" because many incidents reported may be of a minor nature not requiring immediate attention from management.

The top part of the form is filled in by the QA Director, describing the irregularity as he perceives it, and using attached photocopies or the like to support his case. The form is given to the laboratory supervisor to complete. He must investigate the problem, assign a probable cause, and describe efforts made to eliminate the problem. He returns the form to the QA Director who circulates it to top management for initialing indicating that they have received and read the report. The QA Director then files the report and follows up to see that the problem has been corrected. This mechanism forces action on the part of the laboratory supervisor, ensures that top management has been informed, and provides documentation that they have been informed and that action has been taken.

WRITING THE QUALITY ASSURANCE MANUAL

For many laboratories, the written Standard Operating Procedures which define the quality assurance program will be sufficient docu-

No. __________

Quality Assurance Irregularity Report

Part I (To be filled out by QA Director)

1) Date:
2) Sample number(s) involved:
3) Nature of QA irregularity:

Signed ____________________
Q. A. Director

Part II

1) Steps taken to investigate irregularity:

2) Explanation of probable cause irregularity:

3) Steps taken to prevent future occurrence:

4) Name of analyst who performed work:

5) Signed ____________________ Date ______________

Figure 8-2 Quality Assurance Irregularity Report Form

mentation. However, other laboratories, e.g., those seeking accreditation by an outside organization, or those seeking government contracts, or laboratories with discriminating clients, will find that they are increasingly being asked to produce a "Quality Assurance Manual." In general, the document being sought is somewhat broader in scope than the set of SOP's, since it will be concerned with company policy, table of organization of the laboratory, and other topics not covered by the SOP's.

Many organizations have written guidelines to be used in writing a quality assurance manual. The following are those based on the recommendations of Task Force D of the International Laboratory Accreditation Conference (ILAC), of which the author is a member. It should be noted that these recommendations are written rather broadly since they are intended to cover all types of testing laboratories, not limited to analytical chemistry. Some of the items listed may not be pertinent to analytical chemistry, while others perhaps should be expanded.

TYPICAL CONTENTS OF A QUALITY MANUAL FOR TESTING LABORATORIES

1. TABLE OF CONTENTS

2. QUALITY POLICY

2.1 *Objective*

Present a statement of the goals which are expected to be achieved by implementation of the provisions of the Quality System. These goals are commonly given in a management policy statement on the subject.

2.2 *Resources Employed*

List of resources allocated to implement this policy, such as, human, technical, organizational and physical.

2.3 *Quality Assurance Management*

Identify the staff member(s) responsible for developing, implementing, and updating the Quality Manual, Designate the staff organization responsible for carrying out the laboratory's quality assurance program.

3. DESCRIPTION OF THE QUALITY MANUAL

3.1 *Terminology*

Present definition of terms used in the Quality Manual, whenever applicable, making use of internationally recognized definitions whenever they exist. (See list at end of document)

3.2 *Scope*

Present a brief statement of the overall plan by which the objective is to be achieved.

3.3 *Fields of Testing Activity*

Identify specific testing areas covered by the Quality Manual.

3.4 *Management of the Quality Manual*

Designate the individual responsible for updating, revising and distributing the Quality Manual and the process by which this is accomplished. Designate members of the laboratory staff having access to the Quality Manual.

4. DESCRIPTION OF THE LABORATORY

4.1 *Identification*

Name, address, type of corporate structure, and whether or not an affiliate of a larger organization and any other information needed to identify the laboratory.

4.2 *Fields of Activity*

Features of the laboratory or its operations needed to convey a true picture of the organization, such as location and size of all branch laboratories, types of services offered, major fields of activity, and the like.

4.3 *Organizational Structure*

Organizational chart or diagram showing lines of authority and allocation of functions, including that of the Quality System.

4.4 *Responsibility for the Quality Assurance System*
Describe the lines of responsibility for developing, implementing and updating the Quality Assurance System. Describe the relationship of the Quality Assurance staff to other relevant laboratory staff.

4.5 *Technical Management Personnel*
Person(s) having technical management authority in the area(s) covered by the Quality Manual and the line of authority and communication with the manager, or personnel responsible for operation of the Quality System.

4.6 *Documentation of Employee Responsibility*
Written instructions and information that have been given to members of staff to ensure that each employee is aware of the extent and limitations of his area of responsibility.

4.7 *Deputy Assignments*
Present the management directive(s) assigning responsibility for exercise of the management function for the senior technical staff and for the Quality System should the regularly assigned staff be absent.

4.8 *Minimizing Improper Influence*
Management policies designed to assure quality of laboratory tests by minimizing improper influence which might impact adversely upon actions of the laboratory's personnel.

4.9 *Proprietary Rights and Confidential Information*
Measures which the laboratory employs to protect proprietary rights and confidential information.

5. *STAFF*

5.1 *Job Descriptions*
Provide job descriptions for the senior technical staff members.

5.2 *Personnel Records*
Indicate how records covering the education and technical experience of the testing laboratory staff and quality system are maintained. Where an employee has received special training to perform specific tasks or to use specific pieces of apparatus and this information is not conveyed through the normal education credentials, such information should be noted in the employee's personal records file.

5.3 *Supervision of Personnel*
Provide information on each technical operating unit as to the number of supervisory and non-supervisory personnel and of the measures employed to ensure the adequacy of supervision.

5.4 *Other Measures*
List or describe other features pertaining to the laboratory staff which are designed to enhance quality of the laboratory's work, such as, recruitment policy, in-house training, incentive programs to motivate personnel.

6. *EQUIPMENT, TESTING AND MEASURING**

6.1 *Inventory*

(a) *In-House Equipment*
List the in-house major items of test equipment and measuring instruments required to perform the testing covered by the Quality Manual.

*As used herein, measuring equipment refers to all items requiring calibration.

(b) *Records*

Present the following information for each major item of test and measuring equipment in the form of a record: name of item of equipment; name of manufacturer of equipment; type identification and serial number; date equipment received and date placed in service; current location of equipment in laboratory; and, where appropriate, details of maintenance.

6.2 *Identification of Equipment Subject to Calibration*

Describe the method that is used to identify equipment subject to calibration, as to the date of the last calibration and the due date of the next calibration. Describe the system employed to alert user personnel of the time a recalibration is due.

6.3 *Maintenance*

(a) *Periodic Maintenance*

Present information to indicate the measures that are taken for performing required periodic maintenance or reference should be made as to where such instructions are available in the laboratory.

(b) *Overloaded or Mishandled Equipment*

Present a copy of the management directive(s) instructing the staff as to the procedure which it is to follow when an item of equipment has been overloaded or mishandled.

6.4 *Calibration and Checking*

(a) *Calibration Prior to Use*

Document that measuring and testing equipment used in conduct of testing covered by the Quality Manual is calibrated prior to being placed in service.

(b) *Calibration Programs*

Present a description of the overall program of calibration. Identify outside competent sources of calibration which are used and indicate the line of traceability to national or international standards of measurement. Where in-house calibration programs are utilized, indicate the line of traceability of reference standards to national or international standards of measurement.

(c) *Restricted Use of Reference Standards*

Present the management directive(s) and actions intended to ensure that reference standards of measurement are used only for calibration purposes.

(d) *Checking of In-Service Testing Equipment*

Present the management directive(s) and actions which specify the conditions and frequency at which in-service testing equipment is to be checked between regularly scheduled calibrations.

6.5 *Purchasing and Acceptance Procedures for Equipment Consumables/Expendables*

Outline the precautions taken when purchasing and checking equipment and consumables such as contents of purchase orders, conditions for acceptance (criteria and control methods), required documentation (instructions for use, for maintenance and calibration reports), identification, and the like.

7. *ENVIRONMENT*

Present a brief description of not only how the required environmental conditions in the test area are achieved, but also a description of the building facilities, their location and construction features. Present a brief description of the control system employed where the test method requires monitoring of certain environmental conditions. Describe practices relating to control of access to testing areas and, where necessary, practices relating to good housekeeping. Present information to indicate the measures that are taken to protect equipment from the effects of corrosion and other deteriorating atmospheres or abuses.

8. *TEST METHODS AND PROCEDURES*

8.1 *Index of Testing Documents*
List all the standards, instructions, equipment operating manuals and reference data needed to perform the specific tests, or series of tests, covered by the Quality Manual, and the location of these items within the testing laboratory.

8.2 *Use of Non-Standard Test Methods*
Present a full description of any non-standard test method employed.

8.3 *Selection of Test Methods and Testing Sequence*
Present specific quality assurance procedures that relate to the selection of specific tests, or series of tests, covered by the Quality Manual.

9. *UPDATING AND CONTROL OF DOCUMENTS AFFECTING QUALITY*

Describe the system (procedure) employed to ensure that all instructions, standards, operating manuals, and reference data used in performing tests covered by the Quality Manual and maintained up-to-date and designate locations, within the laboratory, where the staff has ready accessibility to these items.

10. *HANDLING OF SAMPLES/ITEMS TO BE TESTED*

10.1 *Receipt and Disposal*
Describe the system which is used to receive, subsequently identify, and dispose of the sample(s) presented for testing.

10.2 *Protection from Damage*
Describe the laboratory's practice on handling samples to preclude their exposure to injurious effects of contamination, corrosion, or mechanical injury and other types of damage.

10.3 *Security*
Describe the practices relating to use of bonded storage.

11. *VERIFICATION OF RESULTS*

11.1 *Verification of Data*
Describe the techniques which the testing laboratory employs to check and verify its calculations and data transfers.

11.2 *Computerized Data*
Present a brief description of how an electronic data processor, if employed, self-protects against inaccuracies due to malfunctions.

12. *TEST REPORTS*

12.1 *Report Format*
Present a specimen test report(s) to illustrate the form and nature of presenting the test results and supporting documents where applicable.

12.2 *Report Revision*
Present management directives(s) or policy statement pertaining to the handling of corrections or additions to reports already issued.

13. DIAGNOSTIC AND CORRECTIVE ACTIONS

13.1 *Feedback and Corrective Actions*

Describe the procedures or means which the laboratory employs to secure feedback on its operation and the considerations that are given to such information, and how corrective actions, if necessary, are carried out.

13.2 *Proficiency and Interlaboratory Comparison Testing*

Present records and action(s) taken with respect to the laboratory's participation in proficiency testing programs and similar interlaboratory comparison testing.

13.3 *Participation in Interlaboratory Correlation Programs*

Identify interlaboratory correlation programs in which the laboratory participates and reference the most recent series of test results obtained.

13.4 *Use of Reference Materials*

Present information with respect to the laboratory's use of Reference Material.

13.5 *Technical Complaints*

Describe the method(s) by which the laboratory responds to technical complaints.

13.6 *Quality System Audit*

Establish the frequency of periodic reviews of the entire Quality System of the laboratory, designate who is to conduct such reviews and the management person to receive the report of the reviews.

14. RECORDS

14.1 *Maintenance of Records*

Describe how the testing laboratory maintains its records pertaining to original observations, calculations and derived data, calibration and equipment maintenance records and final test reports. This description should indicate where such records are held and for how long.

14.2 *Confidentiality and Security*

Present the management directive or policy pertaining to security and confidentiality of test reports and other records referred to in Section 11.1.

14.3 *Historical File of Test Methods*

List the test methods for which an historical file is maintained.

15. SUBCONTRACTING

15.1 *External Equipment*

Identify those items of equipment which the testing laboratory does not have in-house for conduct of certain required tests. List the name and address of the party(s) who makes such equipment available to the test laboratories. Provide information to verify that the equipment is suitable to meet the requirements of the primary laboratory.

15.2 *Subcontracting/Use of External Facilities*

Describe the practices employed to assure that in those cases where test work is performed on a subcontracting basis or testing is performed by the primary laboratory's personnel in other than its in-house facilities, that the responsibilities and obligations of the primary laboratory, pertaining to the test(s), are fully met.

INTERNATIONALLY RECOGNIZED DEFINITIONS

Include the following internationally recognized definitions:

Testing Laboratory (ISO Guide 2). A laboratory which measures, examines, tests, calibrates or otherwise determines the characteristics or performance of materials of products.

Test Method (ISO Guide 2). A defined technical procedure to determine one or more specified characteristics of a material or product.

Test Report (ISO Guide 2). A document which presents the test results and other information relevant to the test.

Quality System (ISO/176). The organization structure, responsibilities, activities, resources, and events that together provide organized procedures and methods of implementation to insure the capability of the organization to meet quality requirements.

NOTE: The Quality System encompasses all elements of quality assurance and quality control. The term Quality System refers to the "total quality system."

OTHER DEFINITIONS

Include other definitions as follows:

Quality Manual. A Quality Manual of a testing laboratory is a document or set of documents intended to give confidence to the laboratory's work and which indicates the specific methods and procedures by which the laboratory achieves its quality objective.

NOTE: The manual may contain other information and operating procedures of the laboratory to satisfy its own needs.

Additional Definitions. Include additional terms which are used in the Quality Manual.

9
LABORATORY ACCREDITATION

In previous chapters reference has been made to the need for good quality assurance in situations where a laboratory may be seeking "accreditation" for the work that it is doing. Perhaps the best definition of laboratory accreditation is "the verification by a competent, disinterested third party that a laboratory possesses the capability to provide accurate test data and that it can be relied upon in its day-to-day operations to maintain high standards of performance."[1] The terminology "third party" implies that the first two parties are the laboratory and its clients or users of the data it supplies. The emphasis in the definition on accuracy and high standards of performance points up the importance of quality control, and the documentation necessary for third party verification emphasizes the need for quality assurance. Finally, the "competent, disinterested" qualification of the third party implies that this party has no financial interest in the outcome of the verification, i.e., the verification must come from outside the laboratory and its parent organization.

The terms "accreditation" and "certification" may be somewhat confusing. These terms have often been used synonymously in the past, along with other similar words, e.g., "approved," "registered," "authorized," etc. In recent years, due to efforts of international standards organizations, agreement seems to have been reached to limit the term "accreditation" to organizations and "certification" to individuals or products. Thus a laboratory is accredited, while a person may be a "Certified Industrial Hygienist," or "Certified Public Accountant." Similarly, a product may be certified by a laboratory which means that it has been found to comply with certain specifications as determined by standard test methods. The reader is warned, however, that these definitions are not universally agreed upon, and one will still hear of certification of laboratories by organizations, e.g., EPA or AIHA, when accreditation is meant.

NEED FOR LABORATORY ACCREDITATION

Certain professions in our society have had a form of personal certification for many years. One cannot practice medicine, for example, without a license and to obtain a license one needs to have a degree in medicine. Similarly, lawyers must pass their bar examinations in order to practice law. Many other professions have similar requirements for practitioners with all of the force of government authority to prevent unqualified persons from designating themselves as practitioners.

Unfortunately, no such system exists for analytical chemists, or any other type of chemist. In most parts of the world, there is nothing to prevent any person from calling himself a chemist, or setting up a laboratory to run chemical analyses. This is true for many other scientific and engineering professions.

Until the last few decades there was little perceived need for laboratory accreditation, and the need that did exist was satisfied by numerous small-scale accrediting programs. However, as described in previous chapters, society is currently entering a highly technical phase, in which great dependence is being placed on the information generated in analytical chemical and other types of testing laboratories. It is thus becoming more and more imperative to establish mechanisms for verifying the accuracy and/or validity of that information.

The need for laboratory accreditation was first felt by the independent commercial testing laboratories. These laboratories are essentially owned and operated by engineers and scientists in private practice making their living by testing for other commercial or private organizations. The independent laboratories which were dedicated to high quality performance were often underbid by competitors who were less dedicated and less professional in their operations. Unfortunately, the clients or purchasers of these services were often technically naive and unable to distinguish between the capabilities of the two classes of laboratories. Hence, it was perceived that a "third-party verification" of laboratory competence would benefit both the high-quality laboratories and their clients.

The second group to perceive the need for laboratory accreditation was the regulatory government agencies, e.g., EPA, FDA, OSHA, NIOSH, etc. The passage of laws, e.g., the Occupational Safety and Health Act, the Clean Drinking Water Act, the Clean Air Act, Resource Conservation and Recovery Act, and many similar laws mandated a massive increase in the amount of testing of all types. Since government testing facilities were limited, much of the work had to be subcontracted to private testing laboratories, non-profit organizations, or universities. Because of the economic importance of this testing, it wasn't long before certain scandals were unearthed indicating that some of these laboratories were less than honest or professional in reporting results. It became obvious that a third-party verification of laboratory competence would be a desirable thing to have.

Finally, the manufacturing industries which were forced to test for compliance with various regulations often found themselves embroiled in disputes with the regulatory agencies over the validity of the technical data generated within their own laboratories. Once again, the need for third-party verification became apparent.

HISTORY OF LABORATORY ACCREDITATION IN THE USA

As mentioned, until the passage of the various federal acts which established the regulatory agencies, there was not a universally perceived need for laboratory accreditation. However, this does not mean that there were no laboratory accreditation programs prior to that time. These were fragmented and established by various organizations, e.g., professional societies, trade organizations, federal, state, and municipal entities, and even private companies which would "accredit" or "certify" suppliers of goods and services after appropriate evaluations. In 1976, in a study performed for the Department of Commerce, Hyer[2] identified some fifty-six laboratory accreditation programs, although the study was not exhaustive. These were about equally divided between federal, state, and municipal agencies and trade associations. These accreditation programs were generally pretty narrow in scope, focusing on certain areas of professional competence (clinical chemistry, industrial hygiene), or certain products or materials (plywood, construction materials), or the ability of a laboratory to perform certain standardized tests.

It is important to note that there are, in general, two conflicting types of accreditation programs, or perhaps it would be more accurate to say two different philosophies concerning the meaning and mechanism for accomplishing laboratory accreditation.

The first of these has been called accreditation by product/standard and represents those programs which accredit laboratories based on their ability to perform certain standard tests, usually on a specified type or class of product. The second type of accreditation has been called accreditation by scientific discipline or field of testing and represents those programs which accredit laboratories by their competence to perform a range of tests within a defined scientific field, e.g., chemical analysis, on a range of products, or on unspecified products.

The American Council of Independent Laboratories (ACIL) has been a strong advocate of laboratory accreditation since its inception in the mid-thirties. ACIL is a trade association of testing and research laboratories of all types, and its members are pledged to promote and advocate high professional standards. As businessmen, they are also acutely aware of the problem of unprofessional, low-quality competition and the need for a system which would assure their clients of the competence of a laboratory.

ACIL is a strong advocate of accreditation by discipline, or, as it is now called, by "field of testing." The reason is that many of its members are broad-based laboratories performing a wide variety of tests on numerous substrates or products. A typical commercial independent analytical laboratory will perform hundreds of analyses on hundreds of different types of materials. The product/standard type of accreditation is simply too complex and too costly for such a laboratory to benefit thereby. On the other hand, discipline or field of testing accreditation would require substantially less cost and be much less complex to handle. It should be pointed out that the typical accreditation process, whether by product/standard or field of testing, represents a direct cost to the accreditee of several thousand dollars per year for each category of accreditation.

In the early 1970's, ACIL and other interested parties began to exert pressure on the federal government to establish a national program for accrediting laboratories. Other countries had established such organizations and a typical model, often cited, was the Australian National Association of Testing Authorities, or NATA, which had a long and honorable record of over thirty years of accreditation. NATA is a field of testing type of accreditation, with some ten fields of testing, e.g., chemical analysis, biological testing, electrical, thermal, etc.

In 1975, the U.S. Department of Commerce announced plans to establish a National Voluntary Laboratory Accreditation Program which would be based on the discipline approach. In 1976, after ten months of study and public comment, the NVLAP was formally established, but was established on the product/standard basis, based presumably on the comments received.

Needless to say, this announcement was greeted with consternation by ACIL and other advocates of the field of testing approach. Realizing that the federal government was committed to the product/standard system, these organizations decided to establish a non-profit, private sector organization to accredit laboratories by field of testing. Thus, at a meeting in 1978 attended by representatives of forty organizations, including ACIL, the American Association for Laboratory Accreditation (AALA) was formed.[3] AALA is a non-profit organization, funded by donations and accreditation fees which accredits laboratories in fields of testing, similar to the Australian NATA system.

Thus, at the present time there are in the United States two national organizations dedicated to accrediting laboratories engaged in testing activities: NVLAP and AALA. For the analytical chemistry laboratory, there is little doubt that the AALA approach is the most suitable. AALA accreditation in the chemical testing field is subdivided into numerous subdisciplines or "testing technologies" corresponding to analytical techniques, e.g., wet chemistry, atomic absorption spectrophotometry, gas chromatography, etc. Each lab specifies those testing technologies in which it claims competence and is evaluated and accredited accordingly.

AALA and NVLAP are the only two major accrediting organizations which exist solely for the purpose of accrediting laboratories. However, there are many other accrediting bodies which have developed from other organizations which have perceived a need for laboratory accreditation in specific areas of laboratory work. Some of these are listed here:

- *The U.S. Environmental Protection Agency* accredits laboratories for the analysis of drinking water for a limited number of parameters. Such accreditation is either done by various state agencies under EPA supervision, or by EPA itself in those states which have not assumed primacy in enforcing the Clean Drinking Water Act. While no accreditation is given for waste water analysis, laboratories performing these tests are monitored for compliance with EPA procedures and are required to analyze proficiency samples annually under the National Pollution Discharge Elimination System permit-granting process.

- *The U.S. Food and Drug Administration* "registers" laboratories engaged in analysis of foods, drugs, cosmetics, and medical devices. Registration requires adherence to Good Laboratory Practices (GLP's) and the program seems to be mainly addressed to laboratories working in support of applications for approval of new drugs.
- *The American Industrial Hygiene Association* accredits laboratories which are engaged in analytical work in support of industrial hygiene studies, especially work-place air contamination. AIHA is a private-sector professional organization, but it has strong, informal ties to the National Institute of Occupational Safety and Health (NIOSH).
- *The College of American Pathologists* is another private-sector, professional group which accredits hospital laboratories.
- *The American Board of Toxicologists* is a private-sector professional group which accredits toxicology laboratories.
- *The American Society of Crime Laboratory Directors* accredits laboratories doing forensic work.

The above list is by no means exhaustive, but is given to show the broad range of accreditation programs available. The private-sector programs are generally voluntary, but in many cases have achieved almost mandatory force in the particular field of accreditation in that they have been recognized by the legal profession as a desirable qualification for laboratory data presented as evidence in litigation.

INTERNATIONAL ASPECTS OF LABORATORY ACCREDITATION

In the past few decades as international organizations, e.g., the Organization for Economic Development (OECD) and International Standards Organization (ISO) have grappled with the problems of international commerce, the lack of an internationally agreed upon system of accrediting laboratories has become apparent as an obstacle to international trade. Quite simply, if a commodity is traded between two countries, it is generally analyzed or tested in both the importing and exporting country, a duplication of effort which could be eliminated if the laboratories in both countries were demonstrably equivalent in competence.

In 1977, at the initiative of the United States and Denmark, the first meeting of the International Laboratory Accreditation Conference (ILAC) was held in Copenhagen. Since then, the ILAC has met every year to investigate problems of international agreement on laboratory accreditation. ILAC is an informal assembly of government personnel and private-sector organizations with an interest in this field. It has no official status, but acts as a forum for discussion of problems.

The work of ILAC is far from completed. It has published a directory of accrediting organizations, defined and discussed many of the legal problems of international recognition of accreditation, listed basic terms and definitions, and produced guidelines for quality assurance systems for laboratories and for accrediting organizations.[4]

CRITERIA FOR LABORATORY ACCREDITATION

When a laboratory applies to an accrediting organization for accreditation, the following aspects of laboratory operation are used in granting such accreditation:

- *Personnel.* Personnel should have education, training, and experience to enable them to perform the tests for which accreditation is requested. Organization of personnel should be such that authority over the technical operations is clearly defined, and responsibility for quality is clearly defined.
- *Facilities.* Laboratory facilities should be adequate to perform the tests designated. Sufficient work space should be provided, necessary utilities available, and the laboratory environment controlled as necessary for the work to be done.
- *Instrumentation.* Instrumentation appropriate to the range of tests being run should be available. Instrumentation should be well maintained and in good condition.
- *Quality Assurance Program.* A suitable quality control and quality assurance program should be established. In most accred-

itation programs today, a *written* quality assurance manual is required.

To ensure that the above criteria are fulfilled, three mechanisms are commonly used:

- On application for accreditation, a laboratory is provided with a detailed questionnaire as part of the application form. This not only provides information for the laboratory assessment but permanent documentation of the laboratory's capabilities.
- After review of the application, if all seems in order, an on-site inspection of the laboratory is conducted by a team consisting of one or more assessors. The inspection covers all of the criteria listed above and more, and frequently uncovers deficiencies which must be corrected before accreditation is granted. If the deficiencies are relatively minor, provisional accreditation may be granted, pending correction of the deficiencies.
- Most accreditation programs require periodic analysis of proficiency samples to demonstrate the laboratory's competence in the areas for which accreditation is sought. In some programs, accreditation is directly dependent on performance in proficiency sample testing, while in others it is used as a means of discovering weaknesses in the laboratory's operation, which the accrediting agency can assist in correcting. Of course, consistent and repeated poor performance will always result in loss of accreditation.

In all accreditation programs, except some of those mandated by federal or state governments, fees are charged. These fees are generally proportional to the extent of testing for which the laboratory is seeking accreditation, and cover the administrative costs of the program and the fees charged by the laboratory assessors.

All accreditation programs are on-going operations and the accredited laboratory must expect to be asked to periodically update its documentation and to be inspected periodically to verify continuing performance. Inspections are generally repeated every two or three years. Proficiency sample testing may be annual, semi-annual, or quarterly, depending on the program.

BENEFITS OF LABORATORY ACCREDITATION

The economic benefits of laboratory accreditation depend on the laboratory and its place in the organization. Many laboratories must be accredited simply to perform the job they are designed to do. Other laboratories may find they need accreditation simply to satisfy other departments of their organization, e.g., the legal department or their own R & D department.

Regardless of these external considerations, it will be found that the accreditation process has certain internal benefits to the laboratory. Examination by a team of objective, technically competent auditors will often point up weaknesses or vulnerabilities in laboratory operations which may not be readily apparent to the laboratory supervisor or Quality Assurance Director. Correction of these deficiencies will result in better laboratory performance. Achieving accreditation through a rigorous program will also be found to have a beneficial effect on the morale of laboratory workers.

REFERENCES

1. Hess, Earl H., Laboratory Accreditation, Professionalism Reduced to Practice, paper presented at the annual meeting of the Texas Council of Engineering Laboratories, at Fort Worth, Texas: January 28, 1984.
2. Locke, John W., *Laboratory Accreditation - State of the Art in 1979,* NBS Special Publication 951, Washington, D.C.: U.S. Department of Commerce, 1980.
3. Amorosi, Roger J., All about AALA, *ASTM Standardization News,* pp 21–23, November, 1980.
4. Forman, Howard I., NBS Special Publication 951, Washington, D.C.: U.S. Department of Commerce, 1980.

BIBLIOGRAPHY

JOURNAL REFERENCES

Dux, J. P., Quality Assurance in the Analytical Laboratory, *American Laboratory,* 212–216 (July 1983)

Garden, J. S., Mitchell, D. G. and Mills, W. N., Non-Constant Variance Regression Techniques for Calibration-Curve-Based Analyses, *Analytical Chemistry,* 52: 2310–2315 (1980)

Glacer, J. A. et al., Trace Analyses for Wastewater, *Environmental Science and Technology,* 15: 1426–1455 (1981)

Horwitz, W., Good Laboratory Practices in Analytical Chemistry, paper delivered at 17th Annual Eastern Analytical Symposium, New York, N.Y. (1977)

Horwitz, W., Kampes, L. R. and Boyer, K. W., Quality Assurance in the Analysis of Foods for Trace Constituents, *Journal of the Association of Official Analytical Chemists,* 63: 1344–1354 (1980)

Hubaux, and Vos, G., Precision and Detection Limits for Linear Calibration Curves, *Analytical Chemistry,* 42: 849–855 (1970)

Kamzelmyer, J. H., Quality Control for Analytical Methods, *ASTM Standardization News,* 25–28 (October 1977)

Keith, L. H. et al., Principles of Environmental Analysis, *Analytical Chemistry,* 55: 2210–2218 (1983)

Kirchmer, C. J., Quality Control in Water Analysis, *Environmental Science and Technology,* 17: 174A–184A (1983)

MacDougall, D. et al., Guidelines for Data Acquistion and Data Quality Evaluation in Environmental Chemistry, *Analytical Chemistry,* 52: 2242–2249 (1980)

Mandel, J., The Analysis of Interlaboratory Test Data, *ASTM Standardization News,* 17–20 (1977)

Mandel, J. and Linnig, F. C., Study of Accuracy in Chemical Analyses Using Linear Calibration Curves, *Analytical Chemistry,* 29: 743–749 (1959)

Provost, L. P. and Elder, R. S., Interpretation of Percent Recovery Data, *American Laboratory,* 57–63 (1983)

Saltzman, B. E., Yeager, D. W. and Meiners, B. G., Reproducibility and Quality Control in the Analysis of Biological Samples for Lead and Mercury, *American Industrial Hygiene Association Journal,* 44: 163–167 (1983)

Schwartz, L. M., Nonlinear Calibration, *Analytical Chemistry,* 49: 2062–2068 (1977)

BOOKS

Dowdy, S. and Wearden, S., *Statistics for Research,* J. Wiley and Sons, New York (1983)

Garfield, F. M., *Quality Assurance Principles for Analytical Laboratories,* Association of Official Analytical Chemists, Arlington, VA (1984)

Garfield, F. M., Editor, *Optimizing Chemical Laboratory Performance Through Application of Quality Assurance Principles,* Association of Official Analytical Chemists, Arlington, VA (1980)

Inhorn, S. L., Editor, *Quality Assurance Practices for Health Laboratories,* American Public Health Association, Washington, D.C. (1978)

Kateman, G. and Pijpers, F. W., *Quality Control in Analytical Chemistry,* J. Wiley and Sons, New York (1981)

N.B.S. Special Publication 591, Berman, G. A. Editor, *Testing Laboratory Performance: Evaluation and Accreditation,* Government Printing Office, Washington, D.C. (1980)

Youden, W. J. and Steiner, E. H., *Statistical Manual of the AOAC,* Association of Official Analytical Chemists, Washington, D.C. (1975)

INDEX

INDEX